Lecture Notes in Mathematics

A collection of informal reports and seminars
Edited by A. Dold, Heidelberg and B. Eckmann, Zürich

208

Alexander Grothendieck

Collège de France, Paris/France

Jacob P. Murre

Rijksuniversiteit, Leiden/Nederland

The Tame Fundamental Group of a Formal Neighbourhood of a Divisor with Normal Crossings on a Scheme

Springer-Verlag
Berlin · Heidelberg · New York 1971

AMS Subject Classifications (1970): 14-02, 14 A 15, 14 B 20

ISBN 3-540-05499-5 Springer-Verlag Berlin · Heidelberg · New York
ISBN 0-387-05499-5 Springer-Verlag New York · Heidelberg · Berlin

Offsetdruck: Julius Beltz, Hemsbach/Bergstr.

INTRODUCTION

In [5] Mumford studied the fundamental group of a neighbourhood of a normal point on an algebraic surface defined over the complex numbers. His method consists in removing the singularity and to study the fundamental group of the corresponding tubular neighbourhood of the total transform of the singular point. This total transform is a divisor with normal crossings on a non-singular model. Mumford is able to express this fundamental group in terms of the fundamental groups of the irreducible curves, which make up the divisor, and in terms of their "intersection configuration".

It has been pointed out in SGA 2 VIII section 3 that this question is also of interest in the case of "abstract" algebraic geometry. For that reason we take up the analogous problem in the context of schemes. In doing so, we have to work with the "algebraic" fundamental group, i.e., with the topological group, which classifies étale coverings (SGA 1 V). Furthermore we have to take formal neighbourhoods. This leads to the fact that most of our results are formulated in terms of formal schemes which have as Ideal of definition an Ideal defined by a divisor. In the applications these formal schemes turn up as completions along the divisor in question. We have to study such coverings of the formal scheme which are, at worst, ramified over the divisor. However, in case there are residue fields with positive characteristics we cannot expect - at least by the present state of affairs - to get reasonable information unless we restrict ourselves to such coverings for which the ramification is tame. We do not try to explain this notion in the introduction (see 2.2.2 for the definition) but mention only that this notion is similar to the familiar one in usual valuation theory.

Roughly the method is as follows: First we study the fundamental group of a formal neighbourhood of an irreducible part of the divisor It turns out that this group is determined -grosso modo- by the fundamental group of the irreducible divisor itself and its self-intersectionclass on the formal scheme; we get a piece of an exact sequence of homotopy groups (7.3.1) analogous to the well-known sequence for the normal sphere bundle of a sub-manifold in a manifold. Next there is a descent theorem (8.2.7) which pieces the fundamental groups of these neighbourhoods of the irreducible components together to the fundamental group of the neighbourhood of the divisor itself.

In SGA 2 XIII section 3 some questions were raised for the fundamental group of the spectrum of a complete noetherian local ring with algebraically closed residue field from which the closed point is removed. We answer some of these problems in case of a local ring of dimension 2 (problem 3.1 ii, 3,2).

At this point we want to emphasis that the fact that -so far- the inseparable part of the fundamental group is neglected, leads to much weaker results than those obtained by Mumford. One of the main results of Mumford is that the fundamental group in question is trivial if and only if the point is non-singular. However, in our case there is the following example,due to M. Artin. Take a 2-dimensional quadratic cone (and consider the completion of the local ring of the singular point) In characteristic zero there is the covering by the plane and the fundamental group is μ_2. In characteristic two this covering is inseparable; the fundamental group considered in this paper is trivial for this example, but the point is singular.

Outline of the paragraphs. We start with the study of coverings of usual schemes which are tamely ramified over divisors with normal crossings (§2). The case of more complicated divisors seems to be of little intrinsic interest. A central result is theorem 2.3.2, due to Abhyankar, which says that such coverings are locally, from the point of view of the étale topology, of a very special type namely quotients of so called Kummer coverings. For the proof of Abhyankar's theorem we refer to SGA 1 XII. The Kummer coverings are treated in detail in §1. In §3 (resp.§4) we introduce the notion "étale covering" (resp."tamely ramified covering") of a formal scheme. An important result in §4 is theorem 4.3.2 which says that, similar as for étale coverings, infinitesimal lifting of tamely ramified coverings is possible and unique. Section 4.4 gives rather special technical results needed in the subtle proof of 7.2.2. In §5 we start the investigation of the fundamental group of a formal neighbourhood of an irreducible divisor. The first steps are very similar to the theory of the inertia group in valuation theory: we obtain the fundamental group in question as an extension of the fundamental group of the divisor itself by the inertia group. In order to go further we need a comparison of two 2-cohomology classes (theorem 6.3.5); this comparison takes place in the context of the abstract 2-cohomology theory as developed by Giraud [4]. From this comparison theorem we obtain in §7 the above mentioned exact sequence of homotopy groups (7.3.1). Finally, §8 contains the descent theorems 8.2.5 and 8.2.7 and §9 the application mentioned earlier.

We thank Giraud for his help in the proof of 6.3.5.

Notations

We follow the notations and terminology of [EGA] and [SGA]. Furthermore we use some symbols which are more or less standard. For instance if n and m are integers then $n \mid m$ (resp. $n \nmid m$) stands for n divides m (resp. n does not divide m). Also if $f: Y \to S$ is a morphism of schemes and U an open piece of S then $f\big|U: Y\big|U \to U$ means the restriction of f to $f^{-1}(U)$.

CONTENTS

§1. Kummer coverings

1.1. Some remarks on diagonizable groups

1.1.1. Let S be a scheme and M an ordinary abelian group. Consider the $\underline{O}_S$- Algebra $\underline{O}_S[M]$, i.e., the Algebra of the group M with coefficients in $\underline{O}_S$. Put

$$D_S(M) = \operatorname{Spec} \underline{O}_S[M],$$

sometimes we write also D(M) instead of $D_S[M]$ if there is no danger of confusion. The S-scheme $D_S[M]$ is in fact a S-group. According to the general definitions in EGA 0_{III} 8.2 it suffices for this to note that:

i) for a variable S-scheme T the $D(M)(T) = \operatorname{Hom}_S(T, D(M))$ is a group, because

$$(*) \qquad D(M)(T) = \operatorname{Hom}_{\underline{O}_S\text{-Algebras}}(\underline{O}_S[M], f_*(\underline{O}_T)) = \operatorname{Hom}_{\text{groups}}(M, \Gamma(T, \underline{O}_T^*))$$

where $f: T \longrightarrow S$ is the structure map,

ii) for a S-morphism $T_1 \longrightarrow T_2$ the corresponding map

$$D(M)(T_2) \longrightarrow D(M)(T_1)$$

is a group homomorphism.

A S-group isomorphic with a S-group $D_S(M)$ is called a diagonalizable group (see SGA 3 I 4.4). Furthermore a homomorphism of groups $\varphi: M \longrightarrow M'$ determines a homomorphism of S-groups $D(\varphi): D(M') \longrightarrow D(M)$.

1.1.2. The formation of D(M) is compatible with base change, i.e., for $S' \longrightarrow S$ we have $D_{S'}(M) = D_S(M) \times_S S'$, because we have canonically

$$\underline{O}_S[M] \otimes_{\underline{O}_S} \underline{O}_{S'} \xrightarrow{\sim} \underline{O}_{S'}[M].$$

If L is a factorgroup of M, then we have a surjective homomorphism $\underline{O}_S[M] \longrightarrow \underline{O}_S[L]$; therefore D(L) is a closed subgroup of D(M). A subgroup of this type is called a diagonalizable subgroup of D(M).

1.1.3. The S-group $\mu_{\underline{n}}$. Let n_i ($i \in I$) be a finite set of positive integers; for abbreviation we write $\underline{n} = (n_i)_{i\in I}$. Consider the S-scheme

$$\mu_{\underline{n},S} = \operatorname{Spec} \underline{O}_S \,[(U_i)_{i\in I}] \,/\, ((U_i^{n_i} - 1)_{i\in I}).$$

The $\mu_{\underline{n},S}$, or shortly $\mu_{\underline{n}}$ if there is no danger of confusion, is a S-group. The $\underline{O}_S$-Algebra of $\mu_{\underline{n}}$ is denoted by $\underline{A}(\mu_{\underline{n}})$.

For variable S-scheme T we have:

$$(**) \qquad \mu_{\underline{n}}(T) = \operatorname{Hom}_S(T, \mu_{\underline{n}}) = \left\{ \underline{\xi} = (\xi_i);\ \xi_i \in \Gamma(T, \underline{O}_T^*) \text{ with } \xi_i^{n_i} = 1 \right\}$$

and this is a group with multiplication

$$\underline{\xi}' \cdot \underline{\xi}'' = (\xi_i' \cdot \xi_i'')_{i\in I}.$$

In terms of the $\underline{O}_S$-Algebra $\underline{A}(\mu_{\underline{n}})$ we have

$$m\colon \underline{A}(\mu_{\underline{n}}) \longrightarrow \underline{A}(\mu_{\underline{n}}) \otimes_{\underline{O}_S} \underline{A}(\mu_{\underline{n}}),$$

with

$$m(\underline{u}_i) = \underline{u}_i \otimes \underline{u}_i \qquad (i \in I),$$

where $\underline{u}_i$ denotes the class of U_i.

1.1.4. For n_i' with $n_i' = n_i \cdot q_i$, with $q_i \in \mathbb{Z}$ ($i \in I$), write $\underline{n} \mid \underline{n}'$. If $\underline{n} \mid \underline{n}'$ then we have a homomorphism of S-groups

$$\varphi_{\underline{n}\,\underline{n}'} : \mu_{\underline{n}'} \longrightarrow \mu_{\underline{n}}$$

obtained from $\mu_{\underline{n}'}(T) \longrightarrow \mu_{\underline{n}}(T)$ by mapping $\underline{\xi}' \longmapsto \underline{\xi} = (\xi_i'^{\,q_i})$.

In terms of the $\underline{O}_S$-Algebras the $\underline{O}_S$-homomorphism $\underline{A}(\mu_{\underline{n}}) \longrightarrow \underline{A}(\mu_{\underline{n}'})$ is given by $u_i \longmapsto u_i'^{\,q_i}$ ($i \in I$).

1.1.5. The group $\mu_{\underline{n}}$ is a diagonalizable group, namely take

$$\mathbb{M} = \mathbb{Z}_{\underline{n}} = \bigoplus_{i\in I} \mathbb{Z}/n_i\mathbb{Z} \quad \text{(additive groups)}.$$

From the formules (*) in 1.1.1 and (**) in 1.1.3 we see that canonically

$$D(\mathbf{Z}_{\underline{n}})(T) \xrightarrow{\sim} \mu_{\underline{n}}(T),$$

i.e., $D(\mathbf{Z}_{\underline{n}})=\mu_{\underline{n}}$. In terms of the $\underline{O}_S$-Algebras this is expressed by the canonical isomorphism

$$\underline{O}_S[\mathbf{Z}_{\underline{n}}] \xrightarrow{\sim} \underline{A}(\mu_{\underline{n}})$$

given by

$$(o,\ldots,1,\ldots o) \mapsto u_i \qquad (1 \text{ on } i^{th}\text{-place.})$$

Note also that in case I consists of one element, i.e., $\underline{n}=n$, the μ_n is the usual group of n-th roots of unity (see SGA 3 I 4.4).

1.1.6. From the formule (*) in 1.1.1 we see that

$$D(M_1 \oplus M_2) \xrightarrow{\sim} D(M_1) \times_S D(M_2).$$

Let us assume from now on that M is finite. Then M is a direct sum of groups of type $\mathbf{Z}/n\mathbf{Z}$, i.e., D(M) is a product of groups μ_n. If the order of M, and hence each n, is prime to the residue characteristics of S (i.e., prime to the characteristic of k(s) for each $s \in S$) then the D(M) is an étale covering of S because this is true for each μ_n (cf.SGA 1 XI 6.3). We recall that an étale covering $f:X \longrightarrow S$ ("revêtement étale" in French) means that f is finite and étale; note however that such f is not necessarily surjective. One has the following lemma:

Lemma 1.1.7. Let S be a locally noetherian scheme, $f:G \rightarrow S$ a finite and étale S-group. Then we have the following:

i) In case S is connected there exists a connected, étale covering S' of S (necessarily surjective) such that $G_{S'}$ is constant, i.e., $G_{S'} \xrightarrow{\sim} \mathfrak{g}_S$ for a suitable ordinary group $\mathfrak{g}$,

ii) for arbitrary S every point $s \in S$ has an open Zariski neighbourhood U with the property mentioned in i); i.e., there exists a surjective étale covering U' of U such that $G_{U'}$ is constant.

Proof: Since S is locally noetherian the connected components are open, therefore it suffices to prove i). Take a geometric point ξ in S and let $\pi = \pi_1(S, \xi)$ be the fundamental group (in the sense of SGA1 V). G corresponds in the category of π-sets with a finite π-group $\mathfrak{g}$. Since the operation of π on $\mathfrak{g}$ is continuous, ther is a normal open subgroup π' of π which operates trivial on $\mathfrak{g}$. Let S' be the connected, étale couvering of S which corresponds with π', then π' is the fundamental group of S' (SGA 1 V 6.13) and $G_{S'}$ is constant.

Corollary 1.1.8. Let the assumptions be as in 1.1.7 and H a closed subgroup of G which is étale over S. Then we have moreover in i) (resp.ii) of 1.1.7. that $H_{S'} \xrightarrow{\sim} \mathfrak{h}_{S'}$ (resp. similar for U') with $\mathfrak{h}$ a subgroup of $\mathfrak{g}$ and the natural embedding $H_{S'} \rightarrow G_{S'}$ corresponds with the natural embedding $\mathfrak{h} \rightarrow \mathfrak{g}$ (resp. similar for U').

Proof: Again it suffices to prove i). H corresponds in the category of π- sets with a π- subgroup $\mathfrak{h}$ of $\mathfrak{g}$. Since π' operates trivial on $\mathfrak{g}$, it operates trivial on $\mathfrak{h}$; the remaining assertions follow from the equivalence between the category of étale coverings of S' and the category of π'- sets.

1.2. Kummer coverings

1.2.1. Let S be a scheme and $\underline{a} = (a_i)_{i\in I}$ a finite set of sections on S, i.e., $a_i \in \Gamma(S, O_S)$. Let furthermore $\underline{n} = (n_i)_{i\in I}$ be a set of positive integers.

Consider the O_S-Algebra:

$$A_{\underline{n}}^{\underline{a}} = O_S [(T_i)_{i\in I}] / ((T_i^{n_i} - a_i)_{i\in I})$$

and put

$$Z = Z_{\underline{n}}^{\underline{a}} = \operatorname{Spec} A_{\underline{n}}^{\underline{a}}$$

with structure morphism $f: Z \longrightarrow S$. Note that Z is faithfully flat and finite over S.

In the following the class of T_i in $\underline{A}_{\underline{n}}^{\underline{a}}$ is denoted by t_i.

The S-group $\mu_{\underline{n},S}$ (shortly : $\mu_{\underline{n}}$) operates on Z over S. Again by general principles of EGA 0_{III} 8.2 it suffices to see that for a variable S-scheme T the group $\mu_{\underline{n}}(T)$ operates on the set $Z(T) = \mathrm{Hom}_S(T,Z)$ and that this operation is functorial. Now we have if $g: T \to S$:

$$Z_{\underline{n}}^{\underline{a}}(T) = \mathrm{Hom}_{O_S\text{-Algebras}}(\underline{A}_{\underline{n}}^{\underline{a}}, g_*(O_T)) = \{\underline{\tau} = (\tau_i)_{i\in I};\ \tau_i \in \Gamma(T,O_T) \text{ with } \tau_i^{n_i} = a_i\}.$$

Then the operation

$$Z_{\underline{n}}^{\underline{a}}(T) \times \mu_{\underline{n}}(T) \longrightarrow Z_{\underline{n}}^{\underline{a}}(T)$$

is given by

$$(\underline{\tau}, \underline{\xi}) \mapsto \underline{\tau}\cdot\underline{\xi} = (\tau_i \cdot \xi_i)_{i\in I}$$

and it is easily checked that this is a group operation and that it is functorial.

Definition 1.2.2. Assume that the a_i are regular (i.e., for every $i\in I$ the a_i is not a zero divisor in the local rings $\underline{O}_{S,s}$ for all $s\in S$). A couple (Y,G) consisting of a S-scheme Y and a S-group scheme G, operating over S on Y, is called a Kummer covering of S, relative to the sections $\underline{a}$, if (Y,G) is isomorphic with a couple $(Z_{\underline{n}}^{\underline{a}}, \mu_{\underline{n}})$ for a suitable set of integers $\underline{n} = (n_i)_{i\in I}$, with each n_i prime to the residue characteristics of S.

One should remark here that a morphism of two couples (Y_i, G_i) $(i = 1,2)$, consisting of S-schemes Y_i and of S-group schemes G_i operating over S on Y_i $(i = 1,2)$, always means a couple (u,φ) with a S-group homomorphism

$$\varphi: G_1 \longrightarrow G_2$$

and a S-morphism

$$u: Y_1 \longrightarrow Y_2$$

compatible with φ.

It follows immediately that the notion of Kummer covering is stable with respect to arbitrary base change provided the inverse images of the sections are regular.

1.2.3. In terms of the O_S-Algebra the action of $\mu_{\underline{n}}$ on $Z_{\underline{n}}^{\underline{a}}$ is given by

$$\phi : \underline{A}_{\underline{n}}^{\underline{a}} \to \underline{A}(\mu_{\underline{n}}) \otimes \underline{A}_{\underline{n}}^{\underline{a}} \quad ,$$

with

$$\phi(t_i) = u_i \otimes t_i \qquad (i \in I).$$

Next we consider the structure of $\underline{A}_{\underline{n}}^{\underline{a}}$ as $\mu_{\underline{n}}$ - $\underline{O}_S$- Module (cf SGA3 I 4.7).

Take $o \leq \alpha_i < n_i$ $(i \in I)$ integers; by abuse of language we write:

$$\underline{\alpha} = (\alpha_i)_{i \in I} \in \mathbb{Z}_{\underline{n}} \quad .$$

For abbreviation, put

$$t^{\underline{\alpha}} = \prod_{i \in I} t_i^{\alpha_i} \quad .$$

The $t^{\underline{\alpha}}$ are a O_S-base for the $\underline{O}_S$-Module $\underline{A}_{\underline{n}}^{\underline{a}}$. In terms of this base, and writing $\underline{u} = (u_i)_{i \in I} \in \underline{A}(\mu_{\underline{n}})$, we have

(*)
$$\phi(t^{\underline{\alpha}}) = \chi_{\underline{\alpha}}(\underline{u}) \otimes t^{\underline{\alpha}}$$

with

$$\chi_{\underline{\alpha}}(\underline{u}) = \prod_{i \in I} u_i^{\alpha_i} \quad .$$

Lemma 1.2.4. Let $\underline{a} = (a_i)_{i \in I}$ and $\underline{a}' = (a_i')$ be two sets of regular sections on S such that $(a_i)\underline{O}_S = (a_i')\underline{O}_S$ $(i \in I)$. Let $\underline{n} = (n_i)_{i \in I}$ be a set of positive integers prime to the residue characteristics of S. Then there exists an étale, surjective covering S′ of S such that over S we have a $\mu_{\underline{n}}$ - isomorphism

$$(Z_{\underline{n}}^{\underline{a}'})_{S'} \xrightarrow{\sim} (Z_{\underline{n}}^{\underline{a}})_{S'} \quad .$$

Proof: We have $a_i = a'_i c_i$ with $c_i \in \Gamma(S, \underline{O}_S^*)$. Take for S' the spectrum of the $\underline{O}_S$- Algebra

$$\underline{O}_S [(V_i)_{i\in I}] / ((V_i^{n_i} - c_i)_{i\in I})$$

and denote by v_i the class of V_i. Clearly due to the assumption on $\underline{n}$ we have that S' is étale over S and the rquired S'- $\mu_{\underline{n}}$ isomorphism is obtained from the isomorphism of $\underline{O}_{S'}$ - $\mu_{\underline{n}}$- Algebras

$$\underline{A}_{\underline{n}}^{\underline{a}} \longrightarrow \underline{A}_{\underline{n}}^{\underline{a}'}$$

given by

$$t_i \mapsto v_i t'_i \quad (i\in I).$$

Lemma 1.2.5. Let S, $\underline{a}$ be as before with regular sections a_i. Let $f: Y \longrightarrow S$ be a finite morphism and $\underline{A} = f_*(\underline{O}_Y)$ the corresponding $\underline{O}_S$- Algebra. Then the following conditions are equivalent:

a) Y is a Kummer covering relative to some sections $\underline{a}'$ with $a'_i = v_i a_i$ and $v_i \in \Gamma(S, \underline{O}_S^*)$ $(i\in I)$.(Note: by this we mean that for some $\underline{n}$ the $\mu_{\underline{n}}$ operates on Y and that $(Y, \mu_{\underline{n}})$ is a Kummer covering).

b) There exists a set $\underline{n} = (n_i)_{i\in I}$ of positive integers n_i prime to the residue characteristics of S such that $\underline{A}$ is a $\underline{O}_S$- Algebra with graduation of type $\mathbb{Z}_{\underline{n}}$ (see SGA 3 I 4.7.3), i.e.,

$$\underline{A} = \coprod_{\underline{\alpha}\in \mathbb{Z}_{\underline{n}}} \underline{A}_{\underline{\alpha}} \quad ,$$

with $\underline{\alpha}$ as in 1.2.3 and such that the $\underline{A}_{\underline{\alpha}}$ have the following properties. Write

$$\underline{A}_{(0,\ldots,1_i,0,\ldots,0)} = \underline{A}_i \quad ,$$

then:

i) $\underline{A}_i$ is free of rank 1,

ii) the canonical map

$$\bigotimes_{i\in I} \underline{A}_i^{\otimes\alpha_i} \longrightarrow \underline{A}_{\underline{\alpha}}$$

is an isomorphism,

iii) the canonical map

$$\rho_i : \underline{A}^{\otimes n_i} \longrightarrow \underline{O}_S$$

gives an isomorphism

$$\underline{A}^{\otimes n_i} \xrightarrow{\sim} (a_i)\, \underline{O}_S .$$

<u>Proof:</u> a) $\Rightarrow$ b) follows from the description in 1.2.3 and the remark $(a_i)\,\underline{O}_S = (a'_i)\,\underline{O}_S$; b) $\Rightarrow$ a) follows after we choose a base t_i in $\underline{A}_i$, we have $t_i^{n_i} = \underline{v}_i\, a_i$ with $v_i \in \Gamma(S, \underline{O}_S^*)$.

<u>1.3. Generalized Kummer coverings</u>

<u>1.3.1.</u> The assumptions are the same as in 1.2.1. Let L be a factorgroup of $\mathbb{Z}_{\underline{n}}$:

$$\mathbb{Z}_{\underline{n}} \longrightarrow L \longrightarrow 0 ,$$

hence

$$K = D(L) \hookrightarrow D(\mathbb{Z}_{\underline{n}}) = \mu_{\underline{n}} ,$$

i.e., $D(L) = K$ is a diagonizable subgroup of $\mu_{\underline{n}}$. Our purpose is to determine the quotientspace $Z_{\underline{n}}^{\underline{a}} / K$. In order to do this we introduce the kernel N as follows:

$$0 \rightarrow N \rightarrow \mathbb{Z}_{\underline{n}} \rightarrow L \rightarrow 0 .$$

By the identification of $\underline{O}_S[\mathbb{Z}_{\underline{n}}]$ with $\underline{A}(\mu_{\underline{n}})$ in 1.1.5 we have that $\underline{\alpha} \in \mathbb{Z}_{\underline{n}}$ corresponds with

$$u^{\underline{\alpha}} = \prod_{i \in I} u_i^{\alpha_i} .$$

Denote the images of u_i in $\underline{O}_S[L]$ by u'_i , then we see that

$$(*) \qquad N = \left\{ \underline{\alpha} \; ; \; \underline{\alpha} \in \mathbb{Z}_{\underline{n}} \text{ with } \prod_{i \in I} {u'_i}^{\alpha_i} = 1 \right\} .$$

N is called the <u>orthogonal</u> of $K = D(L)$. Consider the $\underline{O}_S$- Module

$$\underline{B} = \bigoplus_{\alpha \in N} \underline{t}^{\alpha} \underline{O}_S ;$$

this is in fact an $\underline{O}_S$ -sub-Algebra of $\underline{A}_{\underline{n}}^{\underline{a}}$.

Proposition 1.3.2. We have $\underline{\mu}_{\underline{n}} / K \xrightarrow{\sim} D(N)$ and $\underline{Z}_{\underline{n}}^{a} / K = \operatorname{Spec} \underline{B}$.

Consequently: these quotients are flat over S and commute with arbitrary base change.

Proof: Consider the case $\underline{Z}_{\underline{n}}^{a} / K$; the case $\underline{\mu}_{\underline{n}} / K$ is analogous. First note that the existence of these quotients is guaranteed since the group scheme is flat and the space is affine over the base S (SGA 3 V 4). Furthermore it is well-known that $\underline{Z}_{\underline{n}}^{a} / K = \operatorname{Spec} \underline{B}'$ where $\underline{B}'$ is the subsheaf of $\underline{A}_{\underline{n}}^{a}$ consisting of the invariants under the group action, i.e., the kernel of the couple

$$\underline{A}_{\underline{n}}^{a} \overset{p_i}{\rightrightarrows} \underline{O}_S[L] \otimes_{\underline{O}_S} \underline{A}_{\underline{n}}^{a}$$

with $p_1(\underline{t}^{\alpha}) = \underline{u}'^{\alpha} \otimes \underline{t}^{\alpha}$ and $p_2(\underline{t}^{\alpha}) = 1 \otimes \underline{t}^{\alpha}$ (see 1.2.3). By the formula (*) of 1.3.1 we have indeed $\underline{B}' = \underline{B}$. For the case $\underline{\mu}_{\underline{n}} / K$ we replace $\underline{A}_{\underline{n}}^{a}$ by $\underline{A}(\underline{\mu}_{\underline{n}})$, i.e. by $\underline{O}_S[\mathbb{Z}_{\underline{n}}]$; then we find $\underline{O}_S[N]$ instead of $\underline{B}$.

Corollary 1.3.3. D(N) operates on $\underline{Z}_{\underline{n}}^{a} / K$. We have a natural morphism

$$(u,\varphi) : (\underline{Z}_{\underline{n}}^{a}, \underline{\mu}_{\underline{n}}) \rightarrow (\underline{Z}_{\underline{n}}^{a} / K, D(N)).$$

Proof: The fact that

$$\phi : \underline{A}_{\underline{n}}^{a} \rightarrow \underline{O}_S[\mathbb{Z}_{\underline{n}}] \otimes_{\underline{O}_S} \underline{A}_{\underline{n}}^{a}$$

(see 1.2.3) restricted to $\underline{B}$ factors through $\underline{O}_S[N] \otimes_{\underline{O}_S} \underline{B}$ follows immediately from the formula (*) in 1.2.3 and from the definition of $\underline{B}$. From this factorization follows the corollary at once.

In the same way we get also the following:

Corollary 1.3.4. Let L_1 be a quotient of $\mathbb{Z}_{\underline{n}}$ and L a quotient of L_1:

$$\mathbb{Z}_{\underline{n}} \rightarrow L_1 \rightarrow L .$$

Put $K = D(L)$, $K_1 = D(L_1)$ and let N (resp. N_1) be the kernel of $\mathbb{Z}_{\underline{n}} \rightarrow L$ (resp. $\mathbb{Z}_{\underline{n}} \rightarrow L_1$). Then K is a closed subgroup of K_1 :

$$K= D(L) \hookrightarrow K_1 = D(L_1) \hookrightarrow D(\mathbb{Z}_{\underline{n}}) = \mu_{\underline{n}}$$

and we have a natural morphism

$$(Z_{\underline{n}}^{\underline{a}} / K, D(N)) \longrightarrow (Z_{\underline{n}}^{\underline{a}} / K_1, D(N_1))$$

which induces a natural isomorphism

$$((Z_{\underline{n}}^{\underline{a}} / K) / (K_1 / K), (\mu_{\underline{n}} / K) / (K_1 / K)) \xrightarrow{\sim} (Z_{\underline{n}}^{\underline{a}} / K_1, \mu_{\underline{n}} / K_1 = D(N_1)).$$

<u>Proposition 1.3.5</u>. Let $\underline{a} = (a_i)_{i\in I}$ (resp. $\underline{a}' = (a'_i)_{i\in I}$) be a set of sections and $\underline{n} = (n_i)_{i\in I}$ (resp. $\underline{n}' = (n'_i)_{i\in I}$) a set of positive integers such that $n'_i = n_i . q_i$ with $q_i \in \mathbb{Z}$ and $a_i = b_i^{ni} . a'_i$ with $b_i \in \Gamma(S, O_S^*)$ for $i\in I$. Let furhermore L be a quotient of $\mathbb{Z}_{\underline{n}}$ and put $K = D(L)$. Then there is a morphism of couples:

$$(v, \varphi_{\underline{nn}'}): (Z_{\underline{n}'}^{\underline{a}'}, \mu_{\underline{n}'}) \rightarrow (Z_{\underline{n}}^{\underline{a}}, \mu_{\underline{n}}) ,$$

where $\varphi_{\underline{nn}'}$ is the homomorphism of 1.1.4 , and $(v, \varphi_{\underline{nn}'})$ induces after quotient formation an <u>isomorphism</u> of the couples

$$(Z_{\underline{n}'}^{\underline{a}'} / K', D(N)) \xrightarrow{\sim} (Z_{\underline{n}}^{\underline{a}} / K, D(N)) ,$$

where $K' = \varphi_{\underline{nn}'}^{-1}(K)$ and N is the orthogonal of K (see 1.3.1).

<u>Proof:</u> Consider the exact sequence

$$0 \rightarrow \mathbb{Z}_{\underline{n}} \xrightarrow{\underline{q}} \mathbb{Z}_{\underline{n}'} \xrightarrow{can} \mathbb{Z}_{\underline{q}} \rightarrow 0$$

where $\underline{q}$ is the homomorphism

$$\underline{\alpha} = (\alpha_i) \mapsto \underline{q}.\underline{\alpha} = (q_i \alpha_i) .$$

With the usual identifications $D(\mathbb{Z}_{\underline{n}}) = \mu_{\underline{n}}$, etc. we have that $\underline{q}$ corresponds with the homomorphism $\varphi_{\underline{nn}'}$ from 1.1.4 as follows from the description of $\varphi_{\underline{nn}'}$ on the $\underline{O}_S$-Algebras given in 1.1.4. Consider on the other hand the homomorphism of $\underline{O}_S$-Algebras

$$v: \underline{A}_{\underline{n}}^{\underline{a}} \longrightarrow \underline{A}_{\underline{n}'}^{\underline{a}'}$$

given by

$$v:\ t_i \mapsto \underline{b}_i t_i'^{\,q_i}\ .$$

This identifies $\underline{A}^{\underline{a}}_{\underline{n}}$ with a sub-Algebra of $\underline{A}^{\underline{a}'}_{\underline{n}'}$. Furthermore it is easily checked that v is compatible with the action of $\mu_{\underline{n}}$, resp. $\mu_{\underline{n}'}$ and the homomorphism $\varphi_{\underline{nn}'}$. This proves the first assertion. Next let N' be the image of N under $\underline{q}$

$$\begin{array}{ccccccccc} o & \rightarrow & N & \rightarrow & \mathbb{Z}_{\underline{n}} & \rightarrow & L & \rightarrow & o \\ & & \downarrow\lambda & & \downarrow\underline{q} & & \downarrow\mu & & \\ o & \rightarrow & N' & \rightarrow & \mathbb{Z}_{\underline{n}'} & \rightarrow & L' & \rightarrow & o \end{array}$$

where L' is defined as the quotient. Since λ is an isomorphism we find that K'= D(L') (cf.also SGA 3 VIII 3). Now $N' = \{\underline{q}.\underline{\alpha}\ ;\ \underline{\alpha}\in N\}$ and the $\underline{O}_S$-sub-Algebras of $\underline{A}^{\underline{a}}_{\underline{n}}$ and $\underline{A}^{\underline{a}'}_{\underline{n}'}$ spanned respectively by $\underline{t}^{\underline{\alpha}}$ and $\underline{t}'^{\,\underline{q}.\underline{\alpha}}$ $(\underline{\alpha}\in N)$ correspon with each other by the above identification v. By 1.3.2 these $\underline{O}_S$- sub-Algebras are the $\underline{O}_S$-Algebras of $Z^{\underline{a}}_{\underline{n}}$ / K and $Z^{\underline{a}'}_{\underline{n}'}$ / K' respectively, therefore the last assertion follows.

Corollary 1.3.6. Let $\underline{n}= (n_i)_{i\in I}$ (resp. $\underline{n}' = (n_i')_{i\in I}$) be sets of positive integers with $n_i' = n_i q_i$ $(q_i\in\mathbb{Z})$. Then there is a natural morphism

$$(v,\varphi_{\underline{nn}'}) : (Z^{\underline{a}}_{\underline{n}'},\ \mu_{\underline{n}'}) \rightarrow (Z^{\underline{a}}_{\underline{n}},\ \mu_{\underline{n}})$$

and if K' is the kernel of $\varphi_{\underline{nn}'}$, this induces an isomorphism

$$(Z^{\underline{a}}_{\underline{n}'} / K',\ \mu_{\underline{n}}) \xrightarrow{\sim} (Z^{\underline{a}}_{\underline{n}},\ \mu_{\underline{n}}).$$

Proof: This is a special case of 1.3.5 with $\underline{a}' = \underline{a}$, L= (o).

Corollory 1.3.7. The formation of quotients and the homomorphisms (resp. isomorphisms) in 1.3.2 - 1.3.6 commute with arbitrary base change.

Proof: This follows by looking to the explicite description given.

Definition 1.3.8. Assume that the sections $\underline{a}=(a_i)_{i\in I}$ are regular. A couple (Y,G), consisting of a S-group G and a S-scheme Y on which G operates over S, is called a generalized Kummer covering of S, relative to a given set of sections $\underline{a}$, if ther exists a set of integers $\underline{n}=(n_i)_{i\in I}$, each n_i prime to the residue characteristics of S, and a suitable diagonalizable subgroup K of $\mu_{\underline{n}}$ such that the couple (Y,G) is isomorphic with the couple $(Z_{\underline{n}}^{\underline{a}} / K, D(N))$.

1.3.9. Remarks. a) If (Y,G) is a generalized Kummer covering of S, then it follows from 1.1.6 that G is an étale surjective covering of S and $f:Y \longrightarrow S$ is finite and flat (1.3.2), hence open and closed if S is locally noetherian. Also: it is easily seen that G operates transitively on the fibers.

b) If (Y,G) is a (generalized) Kummer covering of S relative to the sections $\underline{a}$ and if $\varphi: S' \longrightarrow S$ is a morphism such that the inverse images $\underline{a}'=(a'_i)_{i\in I}$ of the $\underline{a}=(a_i)_{i\in I}$ are again regular sections on S' then $(Y_{S'}, G_{S'})$ is a (generalized) Kummer covering of S' relative to the $\underline{a}'$. This follows from 1.3.7.

c) If $(D_i)_{i\in I}$ is a set of divisors on S (note: we consider in this paper only positive divisors, we omit therefore the word positive in the following) and (Y,G) a couple consisting of a S-group G operating over S on the S-scheme Y then we say that (Y,G) is a (generalized) Kummer covering of S relative to the divisors $(D_i)_{i\in I}$, if there exist sections $(a_i)_{i\in I}$ on S such that

$$\operatorname{div}(a_i) = D_i \quad (i\in I)$$

and such that (Y,G) is a (generalized) Kummer covering of S relative to the sections $\underline{a}$.

Lemma 1.3.10. Let $\underline{a}=(a_i)_{i\in I}$ be a set of sections on S and $a_i = \prod_{\lambda\in J_i} a_{i\lambda}$, with $a_{i\lambda}$ sections on S; put $\underline{a}'=(a_{i\lambda})_{i,\lambda}$. Let (Y,G) be a generalized Kummer covering of S relative to the sections $\underline{a}$. Then (Y,G) is also a generalized Kummer covering of S relative to the sections $\underline{a}'$.

Proof: We proceed in two steps.

Case 1: (Y,G) is a Kummer covering $(Z_{\underline{n}}^{\underline{a}}, \mu_{\underline{n}})$. Put $\underline{n}' = (n'_{i,\lambda})_{i,\lambda}$ with $n'_{i,\lambda} = n_i$ ($\lambda \in J_i$, $i \in I$) and consider the subgroup N' of $\mathbb{Z}_{\underline{n}'}$, defined as follows (cf. with 1.2.3 for notations):

$$N' = \{ \underline{\alpha}' = (\alpha'_{i\lambda}) \; ; \; 0 \leq \alpha'_{i,\lambda} < n_i \text{ and } \alpha'_{i\lambda} = \alpha'_{i\mu}, \text{ all } \lambda \text{ and } \mu \in J_i, \text{ all } i \}.$$

Then we claim that there is an isomorphism:

$$(Z_{\underline{n}'}^{\underline{a}'} / D(\mathbb{Z}_{\underline{n}'}/N'), D(N')) \xrightarrow{\sim} (Z_{\underline{n}}^{\underline{a}}, \mu_{\underline{n}}).$$

This isomorphism may be described as follows. Consider the corresponding $\underline{O}_S$- Algebras

$$\underline{A}_{\underline{n}}^{\underline{a}} = \underline{O}_S[\underline{t}], \text{with } \underline{t} = (t_i) \text{ and } t_i^{n_i} = a_i$$

and

$$\underline{B} \subset \underline{O}_S[\underline{t}'], \text{with } \underline{t}' = (t'_{i\lambda}) \text{ and } t'^{\,n_i}_{i\lambda} = a_{i\lambda},$$

with $\underline{B}$ spanned by the elements (see 1.3.1)

$$\underline{t}'^{\underline{\alpha}'} = \prod_{i \in I} \left(\prod_{\lambda} t'^{\,\alpha'_{i\lambda}}_{i\lambda} \right) \text{ with } \underline{\alpha}' \in N .$$

Then consider the $\underline{O}_S$- isomorphism

$$v: \underline{A}_{\underline{n}}^{\underline{a}} \longrightarrow \underline{B}$$

defined by

$$v(t_i) = \prod_{\lambda \in J_i} t'_{i\lambda} .$$

This isomorphism is compatible with the obvious isomorphism

$$\varphi: \mathbb{Z}_{\underline{n}} \longrightarrow N'$$

defined for $\underline{\alpha} \in \mathbb{Z}_{\underline{n}}$ by

$$\Phi(\underline{\alpha}) = \underline{\alpha}' = (\alpha'_{i\lambda}) \quad \text{with} \quad \alpha'_{i\lambda} = \alpha_i \ (i \in I, \lambda \in J_i).$$

The couple (v, $D(\Phi)$) gives the required isomorphism and this completes the proof of case 1.

<u>Case 2 (General case)</u> : Let $(Y,G) = (Z_{\underline{n}}^{\underline{a}} / K, D(N))$ with $N \subset \mathbb{Z}_{\underline{n}}$ a subgroup and $K = D(\mathbb{Z}_{\underline{n}} / N)$. We use the same notations as in case 1; in particular we have $N' \subset \mathbb{Z}_{\underline{n}'}$, and $\Phi : \mathbb{Z}_{\underline{n}} \xrightarrow{\sim} N'$. Consider the following inclusions:

$$N \subset \mathbb{Z}_{\underline{n}} \xrightarrow[\Phi]{\sim} N' \subset \mathbb{Z}_{\underline{n}'} \ .$$

This gives a subgroup N'_1 of $\mathbb{Z}_{\underline{n}'}$ contained in N'. Consider again the isomorphism $(v, D(\Phi))$ of case 1:

$$(Z_{\underline{n}'}^{\underline{a}'} / D(\mathbb{Z}_{\underline{n}'} / N'), D(N')) \xrightarrow{\sim} (Z_{\underline{n}}^{\underline{a}}, \mu_{\underline{n}}) \ .$$

Write $D(\mathbb{Z}_{\underline{n}'} / N') = K'$, $D(\mathbb{Z}_{\underline{n}'} / N'_1) = K'_1$, then the quotient K'_1 / K' corresponds under $D(\Phi)$ with $D(\mathbb{Z}_{\underline{n}} / N) = K$ because N corresponds by Φ with N'_1 . We obtain then - by looking to the Algebras and using 1.3.2 - the following isomorphism

$$((Z_{\underline{n}'}^{\underline{a}'} / K') / (K'_1 / K'), D(N') / (K'_1 / K')) \xrightarrow{\sim} (Z_{\underline{n}}^{\underline{a}} / K, D(N)) \ .$$

However the left-hand side is by 1.3.4 a generalized Kummer covering relative to $\underline{a}'$ and this completes the proof.

Finally we want to determine the automorphisms of a (generalized) Kummer covering.For this purpose we have the following - somewhat more general - result:

<u>Proposition 1.3.11</u>. Let $\underline{a} = (a_i)_{i \in I}$ be a finite set of <u>regular</u> sections on S and let $(Z_{\underline{n}}^{\underline{a}} / K, D(N))$ be a couple as in 1.3.3. Then we have that the D(N)- S-morphisms of $Z_{\underline{n}}^{\underline{a}} / K$ correspond one to one with the sections D(N)(S). It follows in particular that these morphisms are in fact <u>automorphisms</u>. Furthermore if λ is such a D(N)-S-morphism

and if we use the base $\underline{t}^{\underline{\alpha}}$ $(\underline{\alpha} \in \mathbb{N})$ introduced in 1.3.1 then we have for the corresponding $\underline{O}_S$-Algebra homomorphism (denoted by the same letter λ) :

$$\lambda(\underline{t}^{\underline{\alpha}}) = w_{\underline{\alpha}} \cdot \underline{t}^{\underline{\alpha}} \quad (\underline{\alpha} \in \mathbb{N}) \ ,$$

with

$$w_{\underline{\alpha}} \in \Gamma(S, \underline{O}_S^{*})$$

and

$$w_{\underline{\alpha}} \cdot w_{\underline{\beta}} = w_{\underline{\alpha}+\underline{\beta}} \quad (\underline{\alpha}, \underline{\beta} \in \mathbb{N})$$

Proof: Let λ be a D(N)-S-morphism of $Z_{\underline{n}}^{\underline{a}} / K$, then λ is also a $\mu_{\underline{n}}$-S-morphism. First we use only the $\mu_{\underline{n}}$-S-Module structure of $\underline{B}$ (with the notations of 1.3.1). We have

$$\lambda(\underline{t}^{\underline{\alpha}}) = \sum_{\underline{\beta} \in \mathbb{N}} w_{\underline{\alpha}\underline{\beta}} \ \underline{t}^{\underline{\beta}} \qquad (w_{\underline{\alpha}\underline{\beta}} \in \Gamma(S, \underline{O}_S)) \ .$$

Since λ is a $\mu_{\underline{n}}$-homomorphism it has to commute with

$$\phi : \underline{B} \longrightarrow \underline{A}(\mu_{\underline{n}}) \otimes_{\underline{O}_S} \underline{B}$$

given by (we use the notations from 1.2.3):

$$\phi(\underline{t}^{\underline{\alpha}}) = \chi_{\underline{\alpha}}(\underline{u}) \otimes \underline{t}^{\underline{\alpha}} \qquad (\underline{\alpha} \in \mathbb{N}) \ .$$

Using the fact that

$$\phi \ . \ \lambda = (1 \otimes \lambda) . \ \phi$$

we find

$$\sum_{\underline{\beta} \in \mathbb{N}} w_{\underline{\alpha}\underline{\beta}} \ \chi_{\underline{\beta}}(\underline{u}) \otimes \underline{t}^{\underline{\beta}} = \sum_{\underline{\beta} \in \mathbb{N}} \chi_{\underline{\alpha}}(\underline{u}) \otimes w_{\underline{\alpha}\underline{\beta}} \ \underline{t}^{\underline{\beta}} \qquad (\underline{\alpha} \in \mathbb{N})$$

with

$$\chi_{\underline{\beta}}(\underline{u}) = \prod_{i \in I} u_i^{\beta_i} \ .$$

Comparing the coefficients of the base $\chi_{\underline{\gamma}}(\underline{u}) \otimes \underline{t}^{\underline{\delta}}$ $(\underline{\gamma} \in \mathbb{Z}_{\underline{n}}, \underline{\delta} \in \mathbb{N})$ of the $\underline{A}(\mu_{\underline{n}}) \otimes \underline{B}$, we see that for $\underline{\beta} \neq \underline{\alpha}$ we have $w_{\underline{\alpha}\underline{\beta}} = 0$.
Writing $w_{\underline{\alpha}}$ instead of $w_{\underline{\alpha}\underline{\alpha}}$ we have therefore

$$(*) \qquad \lambda(\underline{t}^{\underline{\alpha}}) = w_{\underline{\alpha}} \cdot t^{\underline{\alpha}} \qquad (\alpha \in N)$$

with $w_{\underline{\alpha}} \in \Gamma(S, \underline{O}_S)$. From the fact that $\lambda(1) = 1$ we have moreover

$$w_{\underline{o}} = 1 \ .$$

At this point we are going to use also the <u>multiplicative structure</u> of $\underline{B}$. We have

$$\underline{t}^{\underline{\alpha}} \cdot \underline{t}^{\underline{\beta}} = c_{\underline{\alpha}\underline{\beta}} \ t^{\underline{\alpha}+\underline{\beta}}$$

with $c_{\underline{\alpha\beta}}$ a product of the sections a_i and therefore <u>regular</u>. In fact, according to the conventions of 1.2.3 , we have

$$\underline{\alpha} + \underline{\beta} = (\underline{r}_i(\alpha_i, \beta_i))_{i \in I}$$

with

$$r_i(\alpha_i, \beta_i) \equiv \alpha_i + \beta_i \pmod{n_i}$$

and

$$o \leq r_i(\alpha_i, \beta_i) < n_i \ .$$

From

$$\lambda(\underline{t}^{\underline{\alpha}}) \cdot \lambda(\underline{t}^{\underline{\beta}}) = \lambda(\underline{t}^{\underline{\alpha}} \cdot \underline{t}^{\underline{\beta}})$$

we obtain

$$c_{\underline{\alpha\beta}} \ \underline{w}_{\underline{\alpha}} \ \underline{w}_{\underline{\beta}} \ t^{\underline{\alpha}+\underline{\beta}} = c_{\underline{\alpha\beta}} \ \underline{w}_{\underline{\alpha}+\underline{\beta}} \ t^{\underline{\alpha}+\underline{\beta}} \ .$$

Using the fact that the $c_{\underline{\alpha\beta}}$ are regular we obtain

$$(**) \qquad \underline{w}_{\underline{\alpha}} \cdot \underline{w}_{\underline{\beta}} = \underline{w}_{\underline{\alpha}+\underline{\beta}}$$

Putting $\underline{\beta} = -\underline{\alpha}$ and using $\underline{w}_{\underline{o}} = 1$ we see that

$$\underline{w}_{\underline{\alpha}} \in \Gamma(S, \underline{O}_S^*) \ .$$

From (**) we see that the $\underline{w}_{\underline{\alpha}}$, and hence the λ, determine uniquely an element of

$$\mathrm{Hom}_{\mathrm{groups}}(N, \Gamma(S, \underline{O}_S^*)) = D(N)(S) \ .$$

Conversely it is clear that an element of D(N)(S) determines a D(N)-S-morphism λ of $Z_{\underline{n}}^{\underline{a}}/K$ given by the formula (*) above.

1.4. Inertia groups and connected components of generalized Kummer coverings

1.4.1. Let (Y,G) be a generalized Kummer covering of S relative to a set of sections $\underline{a}$. Let $s \in S$ and ξ a geometric point of S over s, i.e., a morphism

$$\xi : \operatorname{Spec} \Omega \longrightarrow S$$

where Ω is a separably closed field containing k(s).
Let $Y_s = Y \times_S \operatorname{Spec} k(s)$ and $Y_\xi = Y \times_S \operatorname{Spec}\Omega$. Take a point $y \in Y_s$ and a point $\eta \in Y_\xi$ over y, i.e., y is the image of η by the morphism $Y_\xi \longrightarrow Y_s$. Consider also $G_\xi = G \times_S \operatorname{Spec} \Omega$.

Definition 1.4.2. The stabilizer of η in G_ξ is called the inertia group of the point y .

Remark: Since Ω is separably closed and G étale over S, we often consider (by abuse of language!) the G_ξ as an ordinary group. Since G operates transitively on the fibres (cf. 1.3.9 a) and since G is abelian it follows that the inertia group of y is independent of the choice of η above y and in fact independent of y itself, i.e., the inertia group depends only upon $s \in S$. Finally note that in case G is constant the definition agrees with the definition in SGA 1 V page 7, as follows from the remarks made there.

1.4.3. Another description of the inertia group. For $s \in S$ put

$$I_s = \{ i \in I \ ; \ s \in V(a_i) \} \quad ,$$

where, as usual, $V(a_i)$ is the set of the points where a_i is not a unit. Furthermore take a fixed isomorphism

$$(Y,G) \longrightarrow (Z_{\underline{n}}^{\underline{a}}/K, \ D(N))$$

with a suitable $\underline{n}$ and K. Put $G_s = G \times_S \operatorname{Spec} k(s)$ and let G_s^o be the (algebraic!) subgroup of G_s generated by the images of μ_{n_i} with $i \in I_s$, under the homomorphism

$$(\mu_{\underline{n}})_s \to D(N)_s \xrightarrow{\sim} G_s .$$

Note that G_s and G_s^0 both are algebraic group-schemes and not ordinary groups.

Consider $G_\xi^0 = G_s^0 \times_{k(s)} \Omega$. Then we have:

Lemma 1.4.4. G_ξ^0 is the inertia group of an arbitrary point $y \in Y_s$. From this it follows, in particular, that G_ξ^0 is independent of the choice of the isomorphism $(Y,G) \xrightarrow{\sim} (Z_{\underline{n}}^{\underline{a}} / K, D(N))$.

Proof: Make the base change ξ: Spec $\Omega \longrightarrow S$. The proposition follows at once from the description of the operation of $(\mu_{\underline{n}})_\xi$ on $(Z_{\underline{n}}^{\underline{a}})_\xi$ (see (*) of 1.2.3)

Corollary 1.4.5. (Y,G) étale over $s \in S \iff G_s^0 = (e)$

In particular: Y is étale over the points of

$$U = S - \bigcup_{i \in I} V(a_i)$$

and $Y|U$ is a $G|U$-torsor (= principal homogenous covering with group $G|U$) over U.

Proof: First note that Y is flat over S (1.3.2); therefore étaleness is in the present case equivalent with non-ramification and for this we can make the base change ξ: Spec $\Omega \longrightarrow S$. If $G_s^0 = (e)$ then we have étaleness for Y by SGA 1 V 2.3 (note that by 1.3.2 and 1.3.4 we have $Y / D(N) = S$). If, on the other hand, $G_s^0 \neq (e)$ then we have by 1.4.4 that the map

$$N \hookrightarrow \mathbb{Z}_{\underline{n}} \xrightarrow{pr_{i_0}} \mathbb{Z}_{n_{i_0}}$$

is not trivial for some i_0. That means that there exists $\underline{\alpha} \in N$ with

$\alpha_{i_0} \neq 0$; however then the $\underline{O}_S$-Algebra $\underline{B}$ from 1.3.2 is ramified at s. Finally the assertion that over U we have a torsor follows easily by looking to the operation of $G = D(N)$ on a geometric fibre in U.

Corollary 1.4.6. Let S be locally noetherian and connected and the a_i non-invertible on S (i.e., $V(a_i) \neq \emptyset$) for $i \in I$. Then a generalized Kummer covering of S, relative to $\underline{a}$, is connected.

Proof: Since a generalized Kummer covering is a quotient of a Kummer covering, we can start with a Kummer covering (Y,G). Let $Y' \subset Y$ be a connected component, then $Y' \longrightarrow S$ is surjective because the image of Y' in S is open and closed (1.3.9). Let G' be the stabilizer of Y' (we can assume that G is constant, otherwise we replace S by an étale surjective covering, cf. 1.1.7). Since G operates transitively on the fibres of Y, it suffices to show that $G' = G$. For this it suffices to show that $\mu_{n_i} \subset G'$ for all $i \in I$. Fix $i \in I$, take $s \in V(a_i)$ which is, by assumption, possible. If $y \in Y_s$ then we have for $g \in \mu_{n_i}$ that $g \cdot y = y$ by 1.4.4. Hence for $g \in \mu_{n_i}$ we have $g \cdot Y' \cap Y' \neq \emptyset$. Since this intersection is open and closed in Y' we have $g \cdot Y' \subset Y'$, i.e., $\mu_{n_i} \subset G$. This completes the proof.

1.5. Extension of the group of operators

1.5.1. Let X be affine over the ground scheme S, H and G étale and finite S-groups and $\varphi : H \longrightarrow G$ a homomorphism of S-groups and let H operate over S on X (on the right say). Consider $X \times_S G$; on this H operates "by the formula" :

$$(x, g).h = (xh, h^{-1} \cdot g)$$

where $h^{-1} \cdot g$ is an abbreviation for $\varphi(h)^{-1} \cdot g$. It is well-known (cf. SGA 3 V 4.1) that under the above assumptions $(X \times_S G)/H$ exists and is finite over S; this quotient is denoted by $X \times^H G$ (see SGA 1 XI page 11).

The S-group G operates on the right of $X \times_S G$ and the operation is compatible with the operation by H. From this and from the fact that G is flat over S follows- after some diagram chasing - that G operates on $X \times^H G$. One says that $X \times^H G$ is obtained from X by extending the group of operators from H to G. (Remark: this terminology is justified by the fact that there is a canonical morphism $\alpha : X \to X \times^H G$, defined "by the formula" $x \mapsto$ class (x,e), compatible with the action of H, where on $X \times^H G$ the action of H is defined via the action of G by means of the homomorphism $\varphi: H \to G$. The existence of α is clear; in order to see that α has the above described property with respect to the group action of H it suffices to look to the functor and to check the compatibility pointset theoretically, which is immediately clear). We state some simple properties:

Lemma 1.5.2. The assumptions are as above. Assume moreover that Ker(φ) is étale (which is automatically fulfilled in i) and ii)). Then:

i) If φ is trivial we have $(X \times^H G) \xrightarrow{\sim} (X / H) \times_S G$.

ii) If $\varphi: H \xrightarrow{\sim} G$ then $X \xrightarrow{\sim} X \times^H G$ (H operates on both sides).

iii) Put $H' = H / \mathrm{Ker}(\varphi)$ then $X \times^H G \xrightarrow{\sim} (X / \mathrm{Ker}(\varphi)) \times^{H'} G$.

iv) If φ is surjective then $X \times^H G \xrightarrow{\sim} X / \mathrm{Ker}(\varphi)$.

v) If we have moreover an étale S-group K and $\psi: G \to K$ then

$$(X \times^H G) \times^G K \xrightarrow{\sim} X \times^H K ,$$

where the right-hand side is defined via the composition $\psi.\varphi: H \to K$. Moreover all the above isomorphisms are canonical and are isomorphisms of spaces with groups of operators.

1.5.3. Some remarks.

First note that Ker(φ) étale over S implies H étale over S (see SGA 3 VI_B, 9.2 vii). Furthermore making if necessary an étale base change, applying 1.1.8 and using the fact that the isomorphisms are canonical, we can assume that $G = \mathfrak{g}_S$, $H = \mathfrak{h}_S$ and φ corresponds with

$\varphi: \mathfrak{h} \to \mathfrak{g}$, with ordinary groups $\mathfrak{g}$ and $\mathfrak{h}$, i.e., we are reduced to constant groups.

Finally we note that we can assume S = Spec A, X= Spec B. Then

$$X \times_S G = \operatorname{Spec} (\bigoplus_{g \in \mathfrak{g}} B)$$

and the operations are now as follows (writing h.b instead of $\varphi(h).b$):

<u>operation of $\mathfrak{h}$:</u>

$$h.(b_g) = (b'_g)$$

with

$$b'_g = h.b_{h^{-1}.g} ,$$

<u>operation of $\mathfrak{g}$:</u>

$$g.(b_{g_*}) = (b'_{g_*})$$

with

$$b'_{g_*} = b_{g_* g} \quad .$$

<u>1.5.4. Proof of 1.5.2.</u>:

i) In case φ is trivial the operation of H on $X \times_S G$ is on X alone and the assertion follows from the fact that quotient formation commutes with étale base change $G \to S$.

ii) From the description in 1.5.3 it is easily seen that the ring of invariants under $\mathfrak{h}$ = H is B and that the operation of $\mathfrak{g}$ on this ring of invariants corresponds with the operation of $\mathfrak{h}$ on B.

iii) Divide first by Ker(φ) and apply i), next divide by $\mathfrak{h}' = \mathfrak{h}/\operatorname{Ker}(\varphi)$.

iv) Apply iii) and next ii) with $H' = G$.

v) From the natural morphism $X \to X \times^H G$ we get a morphism

$$X \times^H K \to (X \times^H G) \times^G K \quad .$$

In order to see that this is an isomorphism we reduce to constant groups as in 1.5.3 and look to the ring of invariants.

<u>Corollary 1.5.5</u> : The assumptions are as in 1.5.2 (with constant groups say) and let $\mathfrak{K}$ be a normal subgroup of $\mathfrak{g}$. Then we have canonically

(as S- $\mathfrak{g}/\mathfrak{k}$ -schemes):

$$(X \times^{\mathfrak{h}} \mathfrak{g}) / \mathfrak{k} \simeq (X / \bar{\varphi}^{1}(\mathfrak{k})) \times^{\mathfrak{h}/\bar{\varphi}^{1}(\mathfrak{k})} (\mathfrak{g}/\mathfrak{k}).$$

In particular we have canonically

$$(X \times^{\mathfrak{h}} \mathfrak{g}) / \mathfrak{g} \simeq X / \mathfrak{h} \ .$$

<u>Proof:</u> By 1.5.2 iv) we have (always canonically)

$$(X \times^{\mathfrak{h}} \mathfrak{g}) \times^{\mathfrak{g}} \mathfrak{g}/\mathfrak{k} \simeq (X \times^{\mathfrak{h}} \mathfrak{g}) / \mathfrak{k} .$$

By 1.5.2 v) and iii) we have

$$(X \times^{\mathfrak{h}} \mathfrak{g}) \times^{\mathfrak{g}} \mathfrak{g}/\mathfrak{k} \simeq X \times^{\mathfrak{h}} (\mathfrak{g}/\mathfrak{k}) \simeq X / \bar{\varphi}^{1}(\mathfrak{k}) \times^{\mathfrak{h}/\bar{\varphi}^{1}(\mathfrak{k})} (\mathfrak{g}/\mathfrak{k}) .$$

We note that there is also:

<u>Corollary 1.5.6.</u> : The assumptions are as in 1.5.2 with constant groups. Then we have as <u>S-schemes</u> (but not canonically, and not as spaces with operators)

$$X \times^{\mathfrak{h}} \mathfrak{g} \simeq \coprod_{\mathfrak{g}/\mathrm{Im}(\varphi)} (X/\mathrm{Ker}(\varphi)) \ .$$

<u>Proof:</u> Look again to the formulas of 1.5.3. It follows then readily that the ring C of $\mathfrak{h}$-invariants may be written as

$$C = \bigoplus_{\mathfrak{g}/\mathrm{Im}(\varphi)} B^{\mathrm{Ker}(\varphi))}$$

where we take right co-sets and the identification of C with this expression depends on a choice of representatives in $\mathfrak{g}/\mathrm{Im}(\varphi)$.

<u>Lemma 1.5.7.</u> Let be given a morphism

$$(u,\varphi) : (X, \mathfrak{h}) \to (Y, \mathfrak{g})$$

with X (resp. Y) an affine S-scheme, on which the constant group $\mathfrak{h}$ (resp. $\mathfrak{g}$) operates. Then there exists canonically a S- $\mathfrak{g}$ -morphism

$$u' : X \times^{\mathfrak{h}} \mathfrak{g} \longrightarrow Y$$

such that the composition of u' with the canonical morphism $\alpha: X \to X \times^{\mathfrak{h}} \mathfrak{g}$ (see 1.5.1) is u.

<u>Proof.</u> Consider the morphisms

$$X \xrightarrow{\alpha} X \times \mathfrak{g} \xrightarrow{v} Y$$

"defined" by the formulas

$$x \mapsto (x,e) \quad \text{and} \quad (x,g) \mapsto u(x).g \quad .$$

Then $u = v.\alpha$. From the fact that u is (by assumption) compatible with the action of $\mathfrak{h}$ (on Y via φ) follows (settheoretically via the functors) that v is invariant under the action of $\mathfrak{h}$ and we obtain u by quotient factorization:

$$\begin{array}{ccc} X \times \mathfrak{g} & \xrightarrow{v} & Y \\ & \searrow \quad X \times^{\mathfrak{h}} \mathfrak{g} \quad \nearrow_{u'} & \end{array}$$

1.6. Generalized Kummer coverings over strict local rings

1.6.1. In this section we are primarily interested in the behaviour of a generalized Kummer covering over a strict local ring. However we take a slightly more general situation. Let $(Y,G) \xrightarrow{\sim} (Z_{\underline{n}}^{\underline{a}}/K, D(N))$ be a generalized Kummer covering over a locally noetherian base S relative to a set of sections $\underline{a} = (a_i)_{i \in I}$. Assume S has the following properties:

i) S is connected.

ii) The S-groups μ_{n_i} are constant for all $i \in I$ (the n_i from $\underline{n}$ in $Z_{\underline{n}}^{\underline{a}}$).

iii) There exists a point $s \in S$ such that if we put

$$I_s = \{ i;\ s \in V(a_i) \}$$

then a_i, with $i \notin I_s$, is invertible on S and the equation

$$T^{n_i} - a_i = 0$$

has a solution in $\underline{O}_S$.

First note that these conditions are satisfied in case S= Spec A with A strict local, if we take for s the closed point. Next note also that (by the usual arguments like in EGA IV §8) if we start with an arbitrary locally noetherian S and $s \in S$ then we can find an étale neighbourhood of s in S and a point over s having the above properties.

Proposition 1.6.2. Let (S,s) have the above properties (for instance: S strict local, s the closed point) with respect to (Y,G). Let I_s

(resp. G^o) be as above (resp. the subgroup generated by the images of the μ_{n_i} with $i \in I_s$). Let Y^o be a connected component of Y. Put furthermore $\underline{a}_s = (a_i)_{i \in I_s}$ and $\underline{n}_s = (n_i)_{i \in I_s}$. Then we have:

$$(Y^o, G^o) \xrightarrow{\sim} (Z_{\underline{n}_s}^{\underline{a}_s} / (\mu_{\underline{n}_s} \cap K), \mu_{\underline{n}_s} / (\mu_{\underline{n}_s} \cap K))$$

and

$$(Y,G) \xrightarrow{\sim} (Y^o \times^{G^o} G, G) \xrightarrow{\sim} (Z_{\underline{n}_s}^{\underline{a}_s} / (\mu_{\underline{n}_s} \cap K) \times^{\mu_{\underline{n}_s}/(\mu_{\underline{n}_s} \cap K)} \mu_{\underline{n}}/K, \mu_{\underline{n}}/K).$$

In particular, (Y^o, G^o) is a generalized Kummer covering of S relative to the sections $\underline{a}_s$.

<u>1.6.3. Remark.</u> In the above definition of G^o we used a fixed isomorphism of $(Y,G) \xrightarrow{\sim} (Z_{\underline{n}}^{\underline{a}} / K, D(N))$. However the G^o is independent of this, since G^o appears as the stabilizer of Y^o as we shall see below.

Also note that $\mu_{\underline{n}}$, D(N), K and G are constant groups due to the assumption ii) and the connectedness of S. There is no harm therefore to "identify" G^o with G_s^o defined in 1.4.3 (by abuse of language, because the first one is a S-group and the other one a k(s)-group!).

<u>1.6.4. Proof of 1.6.2.</u> Take for every $i \notin I_s$ a root of the equation

$$T^{n_i} - a_i = 0$$

This gives a surjective homomorphism $A_{\underline{n}}^{\underline{a}} \to A_{\underline{n}_s}^{\underline{a}_s}$, therefore an embedding

$$Z_{\underline{n}_s}^{\underline{a}_s} \to Z_{\underline{n}}^{\underline{a}} ,$$

which gives (as is easily seen, or- if one wants- by using 1.5.7) a morphism, and in fact an <u>isomorphism</u>

$$(*) \qquad Z_{\underline{n}_s}^{\underline{a}_s} \times^{\mu_{\underline{n}_s}} \mu_{\underline{n}} \xrightarrow{\sim} Z_{\underline{n}}^{\underline{a}} .$$

The formula for (Y,G) in 1.6.2 is obtained- as far as the extreme left and right sides are concerned- by taking the quotient of (*) by K and using 1.5.5 (with $\varphi: \mu_{\underline{n}_s} \to \mu_{\underline{n}}$ the canonical injection).

Furthermore $Z^{\underline{a}_s}_{\underline{n}_s}$ is connected (1.3.6), hence $Z^{\underline{a}_s}_{\underline{n}_s} / (\mu_{\underline{n}_s} \cap K)$ is connected. Since the morphism $Y \to S$ is open and closed and S connected, the connected components of Y meet the fibre Y_s. Since G operates transitively on this fibre, we have that all connected components are isomorphic. From this follows the formula for (Y^o, G^o)(and it is also clear now that G^o is the stabilizer of Y^o).

1.7. Special properties

1.7.1. In this section we study generalized Kummer coverings under the assumption that the base and/or the sections have special properties. We refer to EGA IV 5.7.2 (resp.5.8.2) for the property(S_k)(resp.(R_k)).

Proposition 1.7.2. Let S be a locally noetherian scheme and (Y,G) a generalized Kummer covering of S relative to the sections $\underline{a}=(a_i)_{i\in I}$. Then

i) S has property (S_k) $\Leftrightarrow$ Y has property (S_k).

ii) S has property (R_o) $\Leftrightarrow$ Y has property (R_o).

iii) S is reduced $\Leftrightarrow$ Y is reduced .

Assume moreover that the closed subschemes $V(a_i)$ defined by a_i have no irreducible component in common and are reduced in their maximal points . Then

iv) S has property (R_1) $\Leftrightarrow$ Y has property (R_1).

v) S normal $\Leftrightarrow$ Y normal .

Proof i): $\Leftrightarrow$ by EGA IV 6.4.2 since Y is flat over S and the fibres are zero-dimensional .

ii): Since Y is flat and finite over S we have that a point $y\in Y$ is maximal if, and only if, $s= f(y)$ is maximal and every maximal point of S is of type f(y) with y maximal on Y (f is the structure map $f: Y \to S$). For such a point $y\in Y$ we have (1.4.5.) that Y is étale over S at s. Therefore $\underline{O}_{Y,y}$ is a field iff $\underline{O}_{S,s}$ is a field and this proves ii) by EGA IV 5.8.4.

iii): Follows from i) and ii) since a locally noetherian scheme is reduced iff it has property (R_o) and (S_1) (EGA IV 5.8.5)

iv): $\Leftarrow$ Follows from EGA IV 6.5.3 i) since Y is flat over S.

$\Rightarrow$ Let $f: Y \to S$ be the structure morphism, $y \in Y$ and $s = f(y)$. Suppose $\dim \underline{O}_{Y,y} \leqq 1$. By EGA IV 6.1.3 $\dim \underline{O}_{S,s} \leqq 1$, hence by assumption $\underline{O}_{S,s}$ is regular. If $s \notin V(a_i)$ $(\forall i \in I)$ then Y is étale over S at s (1.4.5), hence $\underline{O}_{Y,y}$ regular. If $s \in V(a_i)$ then $s \notin V(a_j)$ for $j \neq i$ by assumption, furthermore a_i is a local uniformizing parameter t because $V(a_i)$ is reduced. By going to a sufficiently small étale neighbourhood (cf.1.6.1) of s in S we can assume (cf.1.6.2) that Y is obtained from S by means of one equation of the type

$$T^n - t = o \cdot$$

But then $\underline{O}_{Y,y}$ is again regular - as is easily seen - with uniformizing parameter $t^{1/n}$.

v): Follows from i) and iv) because a locally noetherian scheme is normal if, and only if it satisfies (R_1) and (S_2) (EGA IV 5.8.6).

1.8. Divisors with normal crossings

1.8.1. In the following we are primarily concerned with generalized Kummer coverings over a locally noetherian scheme S relative to a set of divisors, (see 1.3.9 c). Moreover we assume that the divisors have "normal crossings". We first recall the definition.

Let S be a locally noetherian scheme and $(D_i)_{i \in I} = \underline{D}$ a finite set of divisors on S. For simplicity we often denote the inverse images of the D_i in Spec $\underline{O}_{S,s} \to S$ by the same letter D_i.

Definition 1.8.2. a) We say that the $(D_i)_{i \in I}$ have strictly normal crossings if for every $s \in \bigcup_{i \in I} \operatorname{supp} D_i$ we have:

i) $\underline{O}_{S,s}$ is a regular local ring ,

ii) if $I_s = \{i;\ s \in \operatorname{supp}(D_i)\}$, then for $i \in I_s$ we have

$$D_i = \sum_{\lambda} \operatorname{div}(x_{i,\lambda})$$

with $x_{i,\lambda} \in \underline{O}_{S,s}$ and $(x_{i,\lambda})_{i,\lambda}$ part of a <u>regular system of parameters</u> at s.

b) We say that the set $(D_i)_{i \in I}$ has <u>normal crossings</u> if for every $s \in \bigcup_{i \in I} \operatorname{supp} D_i$ there exists an étale neighbourhood $S' \to S$ of s in S such that the family of inverse images of the $(D_i)_{i \in I}$ on S' has strictly normal crossings.

Remark: The concept of (strictly) normal crossings is stable by étale base change; Also: in order to check whether a set of divisors has normal crossings, it suffices to do this after an étale base change.

<u>1.8.3. Regular divisor</u>.Let D be a divisor on S. Consider the closed subscheme of S determined by D (see EGA IV 21.2.12); denote this closed subscheme by the same letter D.

<u>Definition</u>. The divisor D is said to be <u>regular at</u> $s \in \operatorname{supp}(D)$ if the subscheme D is regular at s, i.e., if $\underline{O}_{D,s}$ is a regular local ring. The divisor D is called <u>regular</u> if it is regular everywhere.

The above notion is stable by étale base change (because regularity of a local ring is stable by étale base change). Note that by EGA 0_{IV} 17.1.8 we have for D= div(t): D is regular at s $\Longleftrightarrow$ S is regular at s and $t \notin \underline{m}^2_{S,s}$.

<u>Lemma 1.8.4</u>. If $D = (D_i)_{i \in I}$ is a set of <u>regular</u> divisors with <u>normal crossings</u> then D is a set of divisors with <u>strictly normal crossings</u>. In fact if $D_i = \operatorname{div}(x_i)$ then, after dropping the x_i which are units at s, we have that $(x_i)_{i \in I}$ is part of a regular system of parameters at s.

<u>Proof</u>: Let $D_i = \operatorname{div}(x_i)$ at s $(i \in I)$. There exists, by assumption, an étale neighbourhood S' of s in S and a point $s' \in S'$ above s such that

$$D_{i,S'} = \sum_{\lambda} \operatorname{div}(x'_{i,\lambda}) \quad ,$$

in $\underline{O}_{S',s'}$, i.e.,we have $x_i = \prod_{\lambda} x'_{i,\lambda}$

and $(x'_{i,\lambda})_{i,\lambda}$ part of a regular system of parameters at

s' on S' above s on S. However $x_i \in \underline{m}_{S',s'}$ and $x_i \notin \underline{m}^2_{S',s'}$ by the regularity of the divisor $D_{i,S'}$. Hence (x_i) is a part of a regular system of parameters in $\underline{O}_{S',s'}$, hence also in $\underline{O}_{S,s}$ itself.

Proposition 1.8.5. Let $\underline{D} = (D_i)$ be a set of divisors on a locally noetherian, normal scheme S and (Y,G) a generalized Kummer covering of S relative to $\underline{D}$. Then:

i) if the $(D_i)_{i \in I}$ have normal crossings Y is normal,

ii) if the $(D_i)_{i \in I}$ are regular divisors with normal crossings and if (Y,G) is a Kummer covering then Y is regular above the points of $\bigcup_i \mathrm{supp}(D_i)$,

iii) if the $(D_i)_{i \in I}$ are regular divisors with normal crossings and if (Y,G) is a generalized Kummer covering then Y is regular above the points of

$$\mathrm{supp}\, D_i - \bigcup_{j \neq i} \mathrm{supp}\, D_j \qquad (\forall i \in I).$$

Proof: i) This follows from 1.7.2 v.

iii) By EGA 0_{IV} 17.3.3 and 1.3.9 b we are allowed to make an étale base change; by 1.6.2 (and 1.6.1) we can reduce then to the case of a Kummer covering. Therefore iii) reduces to ii).

ii) Follows from lemma 1.8.6 (which is of a more general nature):

Lemma 1.8.6. Let A be a noetherian local ring and $x_1, \ldots, x_k$ a set of elements of A which consists partly of units and the remaining part is part of a regular system of parameters of A. Let $n_i \geq 1$ $(i = 1, \ldots k)$ be integers such that whenever x_i is a unit then n_i is prime to the characteristics of $A/\underline{m}$. ($\underline{m}$ the maximal ideal of A) (Note: the other n_i are arbitrary). Put

$$B = A[T_1 \ldots, T_k] \,/\, (T_1^{n_1} - x_1, \ldots, T_k^{n_k} - x_k).$$

Then B is a regular semi-local ring.

Proof: Proceed in two steps: first adjoin the T_i with x_i a non-unit, next the remaining ones. The second step is étale. Therefore we can restrict ourselves to the first step: assume all x_i are non-units. We want to show more then, namely: B a regular local ring. Furthermore we can disregard the x_i with $n_i = 1$; assume all $n_i \geq 2$. Put

$$C = A[[T_i, \ldots, T_k]] \quad ,$$

C is a regular local ring with maximal ideal $\underline{n} = (\underline{m}, T_i, \ldots, T_k)$. By EGA 0_{IV} 17.1.7 it suffices to prove that the elements

$$\xi_i = T_i^{n_i} - x_i$$

have linearly independent images in $\underline{n}/\underline{n}^2$. In $\underline{m}/m^2$ the images of x_i are linearly independent by assumption; these images are also linearly independent in $\underline{n}/n^2$ because $\underline{m}/\underline{m}^2 \to \underline{n}/\underline{n}^2$ is injective. Since $n_i \geq 2$ the images of ξ_i and of x_i are the same; this completes the proof.

§ 2. Tamely ramified coverings of schemes

2.1. Tamely ramified fields

2.1.1. Let K be a field with a (non-trivial) discrete valuation v. The valuation ring (resp. residue field, resp. value group) is denoted by A_v (resp. k(v), resp. Γ_v). The fact that v is discrete means that Γ_v is isomorphic with $\mathbb{Z}$. The characteristic of k(v) is denoted by p (i.e., p is zero or a prime number).

Note that it is not assumed that K is complete!

Let L denote a finite, separable extension of K. It is well-known that there are only finitely many- inequivalent - extensions of v to L and each of these extensions is again a discrete valuation (see for instance [3], Alg.Comm., chap.6,§8,Th.1 and cor.3 of prop.1 no 1).

Definition 2.1.2. A finite separable extension L of K is said to be tamely ramified over K with respect to v (shortly, if there is no danger of confusion: L is tame over K) if for each extension w of v to L we have

a) if $p \neq o$ then $p \nmid e$, where e is the index $(\Gamma_w : \Gamma_v)$ of Γ_v in Γ_w,

b) k(w) is separable over k(v).

Lemma 2.1.3. In the following L, L' etc., denote fields containing K; if a compositum is considered then it is tacitely assumed that both are contained in an "overfield".

i) Let $L' \supset L \supset K$

$L' \supset K$ tame $\Leftrightarrow$ $L \supset K$ tame and $L' \supset L$ tame with respect to every extension w of v to L.

ii) $L \supset K$ finite, $K' \supset K$ arbitrary but with a discrete valuation v' in K' extending v in k

$$\begin{array}{ccc} L & \text{---} & L.K' = L' \\ | & & | \\ K & \text{---} & K' \end{array}$$

$L \supset K$ tame with respect to $v \implies L' \supset K'$ tame with respect to v

iii) $L_i \supset K$ tame $(i = 1,2) \implies L_1 \cdot L_2 \supset K$ tame.

$$
\begin{array}{ccc}
L_1 & \!\!\!\!-\!\!\!\!- & L_1 \cdot L_2 \\
| & & | \\
K & \!\!\!\!-\!\!\!\!- & L_2
\end{array}
$$

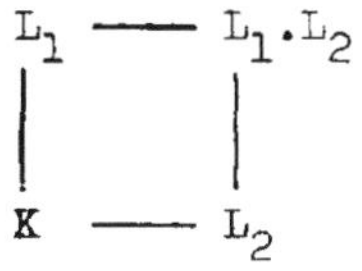

iv) Let $L \supset K$ be tame. Let L_1 be the smallest Galois extension of K containing L. Then $L_1 \supset K$ tame.

v) Let $L \supset K$ be Galois.

$L \supset K$ tame $\iff$ $p \dagger$ order of the inertia group I.

vi) Let $K' \supset K$ be such that there exists a discrete valuation v' in K' extending v with $p \dagger (\Gamma_{v'} : \Gamma_v)$ and $k(v') \supset k(v)$ separable algebraic. Let $L \supset K$ be separable algebraic and such that each of the summands of $L \underset{K}{\otimes} K'$ is tame over K' with respect to v' (shortly: $L \underset{K}{\otimes} K'$ tame over K' with respect to v'). Then $L \supset K$ is tame with respect to v.

$$
\begin{array}{ccc}
L & \!\!\!\!-\!\!\!\!- & L \underset{K}{\otimes} K' \\
| & & | \\
K & \!\!\!\!-\!\!\!\!- & K'
\end{array}
$$

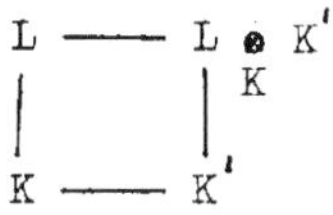

Corollary 2.1.4. Let $\hat{K}$ denote the v- completion of K
$L \supset K$ tame $\iff$ $L \underset{K}{\otimes} \hat{K}$ tame over $\hat{K}$.

2.1.5. Proof of 2.1.3. i) Let w on L extend v and w' on L' extend w. The assertions follow from well-known properties of separable algebraic extensions and from the relation

$$(\Gamma_{w'} : \Gamma_v) = (\Gamma_{w'} : \Gamma_w) \cdot (\Gamma_w : \Gamma_v).$$

ii) Let w' on L' extend v' on K'. Replace L' by the w'-completion $\hat{L'}$, next replace K, K' and L by their closures in $\hat{L'}$. Neither the value groups, nor the residue fields have changed. This reduces therefore the assertion to complete fields; there it is well-known ([7], 3.4.6.; the finiteness condition there is superseded here by the assumption that v' is discrete).

iii) Follows from i) and ii).

iv) Let $L = K(\alpha)$ and $\alpha = \alpha_1, \ldots, \alpha_n$ the conjugates of α in an algebraic closure $\bar{K}$ of K. Clearly each $K(\alpha_i) \supset K$ is tame. Apply iii) to $K(\alpha_1, \ldots, \alpha_n)$.

v) It suffices to consider one extension w to L because L is Galois. Let $e = (\Gamma_w : \Gamma_v)$, $f = (k(w) : k(v))$ and $f = f_i . f_s$ with f_s (resp. f_i) the separable (resp. the purely inseparable) part of f. The assertion then follows from the relation

$$ef_i = \text{order } I.$$

vi) Let A (resp. A') denote the valuation ring in K (resp. K'). Let B (resp. B') denote the integral closure of A(resp. A') in L (resp.in $L \otimes_K K'$) then B' is also the integral closure of $B \otimes_A A'$ in $L \otimes_K K'$ because $B \otimes_A A' \subset L \otimes_K K'$ (because a A-base for B is a K-base for L) and $B \otimes_A A'$ is itself integral over A'. By Cohen- Seidenberg we have that

$$\text{Spec } B' \longrightarrow \text{Spec } B \otimes_A A'$$

is surjective. Since $\text{Spec } (B \otimes_A A') \longrightarrow \text{Spec } B$ is surjective (it is obtained by base change from $\text{Spec } A' \longrightarrow \text{Spec } A$) we have also

$$\text{Spec } B' \longrightarrow \text{Spec } B$$

surjective. This implies, in particular, that every valuation w in L extending v is obtained from a valuation w' in a summand of $L \otimes_K K'$ and extending v'. The assertion follows then from the fact that $k(w)$, as a sub field of the field $k(w')$ which is separable algebraic over $k(v)$, is itself also separable algebraic and from the relation

$$(\Gamma_{w'} : \Gamma_w) . (\Gamma_w : \Gamma_v) = (\Gamma_{w'} : \Gamma_{v'}) . (\Gamma_{v'} : \Gamma_v) .$$

2.2. Tame ramification of normal schemes

2.2.1. In the remaining part of § 2 we make the following assumptions and notations: S is a locally noetherian, normal scheme, D is a closed subset of S of codimension at least one and

$$U = S - D .$$

Note that for $s \in S$ of codimension one (i.e., dim $\underline{O}_{S,s} = 1$) the local ring $\underline{O}_{S,s}$ is a discrete valuation ring. Finally, if X is a scheme then R(X) denotes the function ring of X in the sense of EGA I 7.1.2.

Definition 2.2.2. A morphism $f: X \to S$ (or by abuse of language X itself) is a tamely ramified covering of S relative to the set D if:

1) f is finite,
2) f is étale over U,
3) every irreducible component of X dominates an irreducible component of S,
4) X is normal and
5) for $s \in D$ of codimension 1 in S we have that X is tamely ramified over $\underline{O}_{S,s}$ (see remark 3 below).

2.2.3. Remarks: 1) The use of the word "covering" is somewhat misleading (like the translation of "revetement étale" into "étale covering"), namely f is not necessarily surjective if S is not connected.

2) As to the terminology: instead of "tamely ramified covering of S relative to D" we use also "covering of S tamely ramified over D", or short "X tame over S relative to D", or "X tame over S" if there is no confusion about D.

3) Statement 5 means: Let $X' = X \times_S \mathrm{Spec}\, \underline{O}_{S,s}$ and put $B = \Gamma(X', O_{X'})$. Consider the total ring of quotients of B. By condition 2) this is a direct sum of fields, each of which is a finite separable extension of the quotient field K of $\underline{O}_{S,s}$. Furthermore $\underline{O}_{S,s}$ is a discrete valuation ring in K; denote the corresponding valuation of K by v. Then the requirement is that each of the above summands is tamely ramified over K with respect to the valuation v.

4) The above definition is certainly not the correct one if one wants to study the notion of tame ramification "in full generality". However, here we are primarily interested in the case that D is a

divisor with normal crossings on a normal scheme. In that case the above definition is sufficient as follows - morally - from Abhyankar's theorem. (see 2.3).

2.2.4. Example. Let $\underline{D}=(D_i)_{i\in I}$ be a finite set of divisors with normal crossings on S and (X,G) a generalized Kummer covering of S relative to the divisors $\underline{D}$ (see 1.3.9 c). Put $D=\sum_{i\in I} D_i$, then X is a covering of S tamely ramified with respect to the support of D (or shortly: with respect to D). Condition 1 is immediate, 2 follows from 1.4.5 , 3 from 1.3.9 a and EGA IV 2.3.4 iii, 4 from 1.8.5 and condition 5 from 2.1.3 v and the assumption that the integers n_i, entering in the $\underline{n}$ from $(Z_{\underline{n}}^{\underline{a}}/K, D(N))$, are invertible on S.

Lemma 2.2.5. Let X and Y be normal S-schemes, f: X $\rightarrow$ Y and g: Y $\rightarrow$ S both finite and f surjective. Then

$$\text{X tame over S w.r. to D} \Longleftrightarrow \begin{cases} \text{Y tame over S w.r. to D} \\ \text{X tame over Y w.r. to } D' = g^{-1}(D) \end{cases}$$

Proof: $\Rightarrow$ The fact that Y is étale over U follows from SGA1 V 8.2. (due to the normality of Y and S), next follows by EGA IV 17.7.10 and 17.7.7 the étaleness of X over $g^{-1}(U)$. From the surjectivity of f and the domination of the irreducible components of X over S follows the domination property for the irreducible components of Y. This implies also that D' is at least of co-dimension 1 in Y. From the finiteness of g and the domination property of f.g follows the domination property of f. Condition 5 follows from 2.1.3 i .

$\Leftarrow$ For condition 5 use 2.1.3 i , the other conditions are easy.

Lemma 2.2.6. : Let f: X $\rightarrow$ S be finite, X normal and such that every irreducible component of X dominates an irreducible component of S (and tacitly: S normal). Let X_1 be the normalization of S in the "smallest Galois extension" containing the function ring R(X) of X.

Then: X tame w.r. to $D \Longleftrightarrow X_1$ tame w.r. to D.

Proof: We can assume S connected, hence irreducible. Put $K = R(S)$. Then $R(X) = \bigoplus_{i\in I} L_i$ with L_i separable and finite over K (and I a finite set). Let L_i' be the smallest Galois extension of L_i in the algebraic closure $\bar{K}$. Then X_1 is the normalization of S in $\bigoplus_{i\in I} L_i'$. Clearly we can restrict to the case of one L_i. It is well-known that X_1 is finite over S (see for instance [3], Alg. Comm. Chap. V, Cor.1 of prop.18 § 1 no.6). The étaleness of X_1 over U follows from SGA 1 V 8.2., condition 3 is immediate and condition 5 follows from 2.1.3.iv.

Lemma 2.2.7. Let $f: X \to S$, $\varphi: S' \to S$ étale and surjective. Then: X tame over S w.r. to $D \Longleftrightarrow X' = X_{S'}$ tame over S' w.r. to $D' = \varphi^{-1}(D)$. Note: we assume as always that S is normal, the normality of S' follows from EGA IV 6.5.4.ii).

Proof: First note that for $s' \in S'$ and $s = \varphi(s')$ we have $\dim \underline{O}_{S,s} = \dim \underline{O}_{S',s'}$ (EGA IV 6.1.1.); a similar remark holds for points on X and X'. From this the condition of the domination (condition 3) is readily checked in both directions. Also the finiteness and étaleness are easy.

$\Rightarrow$ The normality of X' follows from EGA IV 6.5.4.ii; condition 5 from 2.1.3.ii.

$\Leftarrow$ The normality of X follows from EGA IV 6.5.4.i; condition 5 from 2.1.3.vi.

Lemma 2.2.8. Given $f: X \to S$ of finite presentation such that $f|U$ is finite and étale. The following conditions are equivalent:

i) X tame over S w.r. to D,

ii) $X \times_S \operatorname{Spec}(\underline{O}_{S,s})$ tame over $\operatorname{Spec} \underline{O}_{S,s}$ w.r. to D $(\forall s \in D)$,

iii) $X \times_S \operatorname{Spec}(\underline{O}^{hs}_{S,s})$ tame over $\operatorname{Spec}(\underline{O}^{hs}_{S,s})$ w.r. to D $(\forall s \in D)$.

(As to the notation: hs means strict localization; in ii) and iii) we mean - of course - the inverse images of D).

Moreover: if we assume f finite, X normal, $f|U$ étale and every irreducible component of X dominates an irreducible component of S then it suffices in ii) and iii) to take points s of codimension 1 in S.

Proof: The normality of $X|f^{-1}(U)$ follows from étaleness (EGA IV 6.5.4ii). ii) $\Leftrightarrow$ iii) as in 2.2.7, with the exception that on uses EGA IV 18.8.12 ii and 6.14.1 instead of EGA IV 6.5.4 in order to prove the normality over $\mathrm{Spec}(\underline{O}^{hs}_{S,s})$.

i) $\Leftrightarrow$ ii) almost immediate from the definitions except for the fact that f finite over the local ring implies f finite in a Zariski neighbourhood (EGA IV 8.10.15).

The last remark in the assertion is immediate since in condition 5 only the points of co-dim 1 enter.

Proposition 2.2.9. Let $f: X \to S$ and $\varphi: S' \to S$ with S' normal and locally noetherian, φ faithfully flat and quasi-compact. Then $X' = X_{S'}$ tame over S' w.r. to $D' = \varphi^{-1}(D) \Rightarrow$ X tame over S w.r. to D.

Proof: From φ faithfully flat we have (EGA IV 6.1.4):

$$\text{co-dim }(D,S) = \text{co-dim }(D',S').$$

From the flatness we have X' normal implies X normal (EGA IV 6.5.4 i), f is finite (by EGA IV 2.7.1) and $f|U$ étale (EGA IV 17.7.1). Also the domination of the irreducible components is easy (EGA IV 2.3.4 iii) Therefore there remains only to be checked condition 5 of definition 2.2.2. Let s (resp. s') be a point of D (resp. D') of co-dimension 1 and $s = \varphi(s')$. Using the last assertion of 2.2.8 (which may be applied in the present situation) we can restrict our attention to $\mathrm{Spec}(\underline{O}^{hs}_{S,s})$. Using the universal property of the strict henselization (EGA 18.8.8 ii) we can replace S' by $\mathrm{Spec}(\underline{O}^{hs}_{S',s'})$; i.e., we can assume S=Spec A, S' =Spec A' with A and A' strict local and discrete valuation rings. Then

$X = \coprod \mathrm{Spec}\, B_\alpha$, with B_α local. It suffices to assume X connected, i.e., $X = \mathrm{Spec}\, B$. Now $f^{-1}(s)$ consists of <u>one</u> point x and $k(x)$ is (at most!) purely inseparable algebraic over $k(s)$. Put $f' = f_{S'}$, then $f'^{-1}(s')$ consists also of one point x' (namely $k(x) \otimes_{k(s)} k(s')$ has only one prime ideal), and $k(x')$ is purely inseparable algebraic over $k(s')$. Since A' is hensel we have X' also connected. Let e denote the ramification index for X, $f = (k(x): k(s))$ and use e', f' similarly for X'. Then we have by the well-known formula for extensions of discrete valuations in the case of separable extensions

$$e.f = (R(X): R(S)) = (R(X'): R(S')) = e'.f' .$$

By assumption p does not divide the right-hand side, hence $p \nmid e$ and $f = 1$.

<u>Lemma 2.2.10.</u> Let $f: X \to S$ be of finite type, D a closed set on S (as in 2.2.1.) and $s \in S$. Assume that X is normal over a Zariski neighbourhood of s and that $X' = X \times_S \mathrm{Spec}\, \underline{O}_{S,s}$ is tame over $\mathrm{Spec}\, \underline{O}_{S,s}$ w.r. to D. Then X is tame over a Zariski neighbourhood of s in S.

<u>Proof:</u> Since our assertion is local on S we can assume that X is normal and, using the tameness over the local ring and EGA IV 8.10.5, that X is finite over S. Furthermore we can assume that S is noetherian and that the irreducible components of X dominate irreducible components of S. Let $T_1 \subset X$ be the set of points where f is <u>not</u> étale, then T_1 is closed. Put $T = f(T_1)$ then, since f is finite, T is also closed. Write $T = D' \cup E$ where D' (resp. E) contains the irreducible components of T contained (resp. not contained) in D. We can assume $s \in T$ since otherwise there is a Zariski neighbourhood of s where f is étale. Now necessarily $s \in D'$, because otherwise the inverse image of E on $\mathrm{Spec}\, \underline{O}_{S,s}$ is not empty and this contradicts condition 2 of the tameness of X'. We get the required neighbourhood of s in S by removing E and those irreducible components of D which do not contain s.

2.3. Tame ramification and Abhyankar's theorem

2.3.1. The following theorem is of central importance. It is due to Abhyankar.

Let- as before- S be a locally noetherian normal scheme, D a divisor on S with normal crossings (1.8.2.). The support of D is denoted by the same letter D and U= S-D.

Theorem 2.3.2. With the assumptions of 2.3.1., let $f: X \to S$ be a finite morphism and $\mathfrak{g}$ an ordinary group operating on X over S. Assume that $X|U$ is a $\mathfrak{g}$-torsor (cf.1.4.5.). Then the following conditions are equivalent:

i) X is tamely ramified over S relative to D,

ii) for every $s \in S$ there exists an étale neighbourhood S' of s in S such that $D_{S'} = \sum_{i \in I} D'_i$, with D'_i divisors on S', and a Kummer covering $(Y, \mathfrak{h})$ of S' relative to the divisors $(D'_i)_{i \in I}$ such that over S' there is an isomorphism.

$$(X_{S'}, \mathfrak{g}) \xrightarrow{\sim} (Y \times^{\mathfrak{h}} \mathfrak{g}, \mathfrak{g}).$$

(Remark: In the above statement $\mathfrak{h}$ is assumed to be an ordinary group; furthermore the statement tacitly implies the existence of a group homomorphism $\mathfrak{h} \to \mathfrak{g}$).

Proof: See SGA I XII.

2.3.3. Remark: It follows from ii) and the remark just preceding 1.2.3 that for "sufficiently small" étale neighbourhood S' of s we have always property ii). Moreover for sufficiently small S' we have the additional information that we can take $(Y, \mathfrak{h})$ to be a Kummer covering relative to divisors $(D'_i)_{i \in I}$ with $D_{S'} = \sum_{i \in I} D'_i$, with $(D'_i)_{i \in I}$ a set of irreducible, regular divisors with normal crossings on S'.

Proof: For S' sufficiently small we have that $D_{S'}$ is a sum of such divisors (see definition 1.8.2), the rest follows from lemma 1.3.10

and the observation that in statement ii) we can always replace a <u>generalized</u> Kummer covering by a Kummer covering (1.5.2).

<u>Corollary 2.3.4</u>. The assumptions on S and D are as in 2.3.1. Let $f: X \to S$ be a morphism of finite type. Equivalent conditions:

i) X tamely ramified over S with respect to D,

ii) for every $s \in S$ there exists an étale neighbourhood S' of s in S such that $X' = X_{S'}$ is a finite disjoint union of generalized Kummer coverings of S' with respect to a set of divisors $(D'_i)_{i\in I}$ with $D_{S'} = \sum_{i\in I} D'_i$. Moreover for sufficiently small S' we can make the same additional assumptions on the D'_i as in 2.3.3. Also if $D'_i \cap D'_j = \emptyset$ $(i \neq j)$ then we can replace in the above statement, <u>generalized</u> Kummer covering by Kummer covering.

<u>Remark</u>: In 2.3.4 we mean, of course, isomorphisms as <u>schemes</u> because we don't have a group action on X itself.

<u>Proof</u>: ii) $\Rightarrow$ i) From 2.2.4 and 2.2.7.

i) $\Rightarrow$ ii) The assertion is local. We can assume S and X irreducible and (EGA IV 8.) S strict local. Let X_1 be the normalization of S in the Galois extension generated by the function field R(X) over R(S) in the algebra closure $\overline{R(S)}$. By 2.2.6 X_1 is tamely ramified over S relative to D. If $\mathfrak{g}$ is the Galois group of $R(X_1)$ over R(S) then we apply 2.3.2 (resp.remark 2.3.3 for the additional information) on $(X_1, \mathfrak{g})$. Since in our case X_1 is irreducible, we have (1.5.2 iii and 1.5.6) that X_1 is itself a generalized Kummer covering. If X corresponds with the subgroup $\mathfrak{h}$ of $\mathfrak{g}$ then we have $X = X_1/\mathfrak{h}$ (because both X and $X_1/\mathfrak{h}$ are normal) and this is a generalized Kummer covering by 1.3.4. This completes the proof except for the remark concerning $D'_i \cap D'_j = \emptyset$ $(i \neq j)$; this remark follows from 1.6.2.

<u>Corollary 2.3.5</u>. The assumptions are as in 2.3.1. If X is tamely

ramified over S relative to D then X is flat over S.

Proof: From 2.3.4 ii), 1.3.9 a and EGA IV 2.5.1.

Corollary 2.3.6. The assumptions are as in 2.3.1. Let $\varphi: S' \to S$, with S' normal and such that $\varphi^{-1}(D) = D'$ is defined and is again a divisor with normal crossings. Let $f: X \to S$ be tame over S relative to D. Then $f' = f_{S'}: X' = X_{S'} \to S'$ is tame over S' relative to D'.

Proof: Over $\varphi^{-1}(U)$ we have étaleness. Let $s' \in D'$ and $s = \varphi(s')$; by 2.2.8 and the universal property of strict henselizations we can replace S' (resp. S) by Spec $\underline{O}^{hs}_{S',s'}$ (resp. Spec $\underline{O}^{hs}_{S,s}$). Therefore we can assume S and S' strict local. Then X is a disjoint union of generalized Kummer coverings (2.3.4); due to the assumption that $\varphi^{-1}(D)$ is defined we have the same for X' (1.3.9 b) and finally, since $\varphi^{-1}(D)$ has normal crossings and (2.3.4), X' is tame relative to D'.

2.4. The category $Rev^D(S)$

2.4.1. The assumptions are as in 2.3.1 , i.e., S is a local noetherian, normal scheme and D a divisor with normal crossings on S. We assume moreover that S is connected (hence irreducible). Let Rev(S) denote the category of S-schemes $f: X \to S$ for which f is finite (i.e., X is a "covering" or "revetement" of S), furthermore RevEt(S) is the category of étale coverings (revetement étale) of S and $Rev^D(S)$ the category of coverings of S tamely ramified relative to D. One has the following inclusions as full sub categories:

$$RevEt(S) \subset Rev^D(S) \subset Rev(S).$$

Theorem 2.4.2. $Rev^D(S)$ is a Galois category (after the choice of a suitable fibre functor, see 2.4.3).

Proof: We have to check the conditions G1,..,G6 of SGA 1 V 4. We use tacitly the corresponding properties for étale coverings.

G1: Existence of a final element: S itself. Existence of products and fibre products:

$$S \leftarrow Z \begin{array}{c} \nearrow X \nwarrow \\ \\ \nwarrow Y \swarrow \end{array} X \times_Z Y \leftarrow (X \times_Z Y)_n$$

Take the normalization of S in the function ring $R(X \times_Z Y)$. This function ring is a finite direct sum of fields, each finite and separable over R(S). Therefore the normalization is finite over S. The remaining requirements for tame ramification are all obvious except the last one, which follows from 2.1.3 iii. Finally the fact that this normalization is the fibre product in $Rev^D(S)$ follows from the universal property of the normalization (EGA II 6.3.9).

G2: Existence of finite sums and quotients. Easy, use 2.2.5 and the fact that the quotient of a normal scheme is normal.

G3: Given $u: X \to Y$ in $Rev^D(S)$, we want a decomposition

$$X \xrightarrow{u'} Y \xrightarrow{u''} Y = Y' \amalg Y''$$

with u' a strict epimorphism and u'' a monomorphism. Take for Y' those components of Y which are dominated by a component of X and for Y'' the remaining ones, then due to the normality of Y we have $Y = Y' \amalg Y''$. The fact that $u' : X' \to Y'$ is an epimorphism follows from the fact that $X|U$ (with $U = S - D$) is schematically dense in X (EGA IV 11.10.2) and from the corresponding property for étale coverings. In order to see that u' is a strict epimorphism we have- if $\varphi . p_1 = \varphi . p_2$ - to complete the diagram:

$$(X \times_{Y'} X)_n \underset{p_i}{\rightrightarrows} X \xrightarrow{u'} Y' \overset{v}{\searrow} Z, \qquad X \xrightarrow{\varphi} Z$$

However over U we have the existence of v. The existence of v over S itself follows, since Z and Y' are normal, from the fact that the R(S) homomorphism $R(Z) \to R(Y')$ induces a S-morphism from the normalization Y' of S in $R(Y')$ to Z (EGA 6.3.9).

2.4.3. The fibre functor. Take a point $s_o \in S$, $s_o \notin D$ and a separably closed field $\Omega \supset k(s_o)$, i.e., a geometric point

$$\xi : \operatorname{Spec}\Omega \to S .$$

Consider the "fibre functor"

$$F(X) = \operatorname{Hom}_S(\operatorname{Spec}\Omega, X) ,$$

where $X \in \operatorname{Rev}^D(S)$. Then G4 and G5 are immediate (see SGA 1 V 4). As to G6: let $u: X \to Y$ be given such that $F(u)$ is an isomorphism; to prove: u is an isomorphism. By the theory of étale coverings we have that $X|U \xrightarrow{u} Y|U$ is an isomorphism. The fact that u itself is an isomorphism follows again from the universal property of normalization (EGA II 6.3.9).

Corollary 2.4.4. There exists a profinite group $\pi_1^D(S,\xi)$ such that the category $\operatorname{Rev}^D(S)$ is equivalent with the category of finite sets on which this group operates continuously (the equivalence is obtained by means of the fibre functor F of 2.4.3). The group $\pi_1^D(S,\xi)$ is called the tame fundamental group of S with respect to D and with base point ξ .

Proof: This follows from 2.4.2 and SGA 1 V 4.

2.4.5. From the results in SGA 1 V follow also the usual properties of the fundamental group. We mention the following:

a) If we change the base point from ξ to ξ' then $\pi_1^D(S,\xi)$ and $\pi_1^D(S,\xi')$ are isomorphic, the isomorphism is determined up to an inner automorphism.

b) If $\varphi: S' \to S$ is as in 2.3.6. and ξ' is a geometric point in S' then there is a continuous homomorphism.

$$\pi_1^{D'}(S',\xi') \to \pi_1^D(S,\xi),$$

determined up to an inner automorphism of $\pi_1^D(S,\xi)$.

c) Galois object in $\operatorname{Rev}^D(S)$. An object X in $\operatorname{Rev}^D(S)$ is called a Galois object with group Γ if Γ operates (on the right, say) on X over S such that:

a) $X \times \Gamma \overset{\sim}{\rightarrow} X \times X$,

b) $X \neq \emptyset$.

Remark: The morphism in a) is the "well-known" morphism given "by the formula" $(x,\gamma) \rightarrow (x,x\gamma)$. Also it would be more correct to write Γ_S (i.e., the constant group Γ over S) instead of Γ.

2.4.6. a) Every X in $Rev^D(S)$ determines a separable finite algebra over the function field R(S) of S. In this way we get (see 2.4.5.b) a continuous, surjective homomorphism

$$(*) \qquad Gal(\overline{R(S)}/R(S)) \rightarrow \pi_1^D(S,\xi) \rightarrow 1$$

where $\overline{R(S)}$ denotes a separable algebraic closure of R(S) and Gal(...) denotes the Galois group. The fact that the homomorphism is surjective is seen as follows: by 2.2.5 the connected components of $X \in Rev^D(S)$ are itself in $Rev^D(S)$, therefore we have:

X connected in $Rev^D(S)$ $\Leftrightarrow$ X connected as scheme (i.e., by normality, X irreducible).

Therefore connected $\pi_1^D(S,\xi)$- sets give connected $Gal(\overline{R(S)}/R(S))$-sets which prove the surjectivity (cf. SGA 1 V 5.3).

b) The kernel of (*) corresponds with the sub field of $\overline{R(S)}$ consisting of the compositum of the finite extension of R(S) in $\overline{R(S)}$ which are "at worst" tamely ramified over S with respect to D (cf.SGA 1 V 8.2).

c) From b) and SGA 1 V 8.2 we see that the homomorphism (*) from a) can be factored into the following continuous homomorphisms (both surjective)

$$Gal(\overline{R(S)}/R(S)) \rightarrow \pi_1(U,\xi) \rightarrow \pi_1^D(S,\xi),$$

where as usual U= S- D.

d) In the same way the continuous surjective homomorphism $Gal(\overline{R(S)}/R(S)) \longrightarrow \pi_1(S,\xi)$ from SGA 1 V 8.2 factors into the following two continuous, surjective homomorphisms (the second one corresponds with the inclusion RevEt(S) $\rightarrow$ $Rev^D(S)$):

$$Gal(\overline{R(S)}/R(S)) \rightarrow \pi_1^D(S,\xi) \rightarrow \pi_1(S,\xi).$$

----

§3. Extension of some notions from the theory of schemes to the theory of formal schemes

3.1. General remarks

3.1.1. In the following $\mathcal{S}$ (resp. $\mathcal{X}, \mathcal{Y}, \ldots$) denotes a formal scheme (formal pre-scheme in the old terminology, EGA I 10.4.2). We assume tacitly that the formal schemes are locally noetherian; this implies that they are adic (EGA I 10.4.2) and that there is a largest Ideal of definition (EGA I 10.5.4).

3.1.2. Comparison between some local rings. Let $\mathcal{S}$ = Spf A with A a J-adic ring; put S= Spec A. Let $s \in \mathcal{S}$; there are canonical local homomorphisms (EGA 0_I.7.6)

$$\underline{O}_{S,s} \xrightarrow{\lambda} \underline{O}_{\mathcal{S},s} \xrightarrow{\mu} \hat{\underline{O}}_{S,s} ,$$

where completion means J-adic completion, the composition $\mu.\lambda$ is the canonical homomorphism from a ring to its completion. Both λ (EGA 0_I 7.6.15 and 6.2.3) and μ (EGA 0_I 7.6.18) are faithfully flat. From this we see by standard arguments (via extension and contraction) that

$$\underline{m}_{S,s} \underline{O}_{\mathcal{S},s} = \underline{m}_{\mathcal{S},s} .$$

Also it is known that the above local rings have the same residue field (EGA 0_I7.6.10 and 7.6.17).

3.1.3. Normal and regular formal schemes. A formal scheme $\mathcal{S}$ is normal (resp. regular) in s if the local ring $\underline{O}_{\mathcal{S},s}$ is normal (resp. regular). Let $\mathcal{S}$ = Spf A be normal in s; it follows from the remarks in 3.1.2 and EGA IV 6.5.2 that this implies normality for S= Spec A in s. Hence normality of $\mathcal{S}$ implies S normal in all closed points, hence normality everywhere

$$\text{Spf A normal} \Longrightarrow \text{Spec A normal} .$$

Similarly $\mathcal{S}$ regular in s implies S regular in s and $\mathcal{S}$ regular implies S regular. Conversely assume S regular in s. From the remarks in 3.1.2 above follows that the condition of EGA 0_{IV} 17.3.3 d is fulfilled; therefore $\mathcal{S}$ is regular in s. Therefore:

$$\mathrm{Spf}\ A \text{ regular} \Longleftrightarrow \mathrm{Spec}\ A \text{ regular} .$$

Moreover we have (EGA 0_{IV} 17.3.3 and 17.1.7): if $t_i \in \underline{O}_{S,s}$ $(i=1,\ldots,r)$ then $(t_1,\ldots,t_r)$ is a regular system of parameters in $\underline{O}_{S,s}$ if and only if $(\lambda(t_1),\ldots,\lambda(t_r))$ is a regular system of parameters in $\underline{O}_{\mathcal{S},s}$.

3.1.4. Divisors with normal crossings. Let D be a closed subset of $\mathcal{S}$. For every $s \in \mathrm{supp}(D)$ the D determines a closed subset D_s of Spec $\underline{O}_{\mathcal{S},s}$. If D is a divisor (always tacitly assumed to be positive!) on $\mathcal{S}$ then D_s is a divisor on Spec $\underline{O}_{\mathcal{S},s}$.

Definition. A divisor D on $\mathcal{S}$ has normal crossings (resp. strictly normal crossings, resp. is regular) if for every $s \in \mathrm{supp}(D)$ the D_s has normal crossings (resp. strictly normal crossings, resp. is regular) on Spec $\underline{O}_{\mathcal{S},s}$ (1.8.2) .

Note: 1) This implies that $\underline{O}_{\mathcal{S},s}$ is a regular local ring; in case of normal crossings this is part of the definition, in case of regularity it follows from EGA 0_{IV} 17.1.8. Moreover in the case of regularity we have $D_s = \mathrm{div}(t)$, with t part of a regular system of parameters.
2) For a family of divisors on $\mathcal{S}$ we give a similar definition.

3.1.5. Next let $\mathcal{S}$ = Spf A, with A a J-adic ring and S= Spec A. Let D be a divisor on $\mathcal{S}$. The corresponding Ideal $\mathcal{J}(D)$ is coherent and determines(EGA I 10.10.5) an ideal I in A, which in turn determines a divisor D (same notation!) on S. In fact the divisor D on $\mathcal{S}$ is the inverse image of this divisor on S by the natural morphism of ringed spaces $\mathcal{S} \to S$. If $\underline{D} = (D_i)_{i \in I}$ is a set of regular divisors with normal crossings on $\mathcal{S}$ then, by 3.1.3 and 3.1.5[1], the corresponding set on S has the same property; the converse statement holds also.

Lemma 3.1.5[1]. Let B be a noetherian, regular local ring, $\underline{p}$ a prime ideal in B and $(t_1,\dots,t_r)$ part of a regular system of parameters in B with $t_i \in \underline{p}$ $(i=1,\dots,r)$. Then $(t_1,\dots,t_r)$ is part of a regular system of parameters in $B_{\underline{p}}$.

Proof: Put $I=(t_1,\dots,t_r)$. By EGA 0_{IV} 17.1.7. the ideal I is prime and B/I is regular. Hence (see EGA 0_{IV} 17.3.2.) $(B/I)_{\underline{p}} = B_{\underline{p}}/I_{\underline{p}}$ is regular, therefore it suffices (again by ibid.17.1.7) to prove that $(t_1,\dots,t_r)$ is part of a system of parameters for $B_{\underline{p}}$. Both B and $B_{\underline{p}}$ are Cohen-Macaulay (17.1.3); applying (ibid 16.5.6, and 16.5.11) to B we see that $\dim(B_I)= r$. Since $B_I = (B_{\underline{p}})_{I_{\underline{p}}}$ we have by (ibid 16.5.11) applied to $B_{\underline{p}}$:

$$\dim(B_{\underline{p}}/I_{\underline{p}}) = \dim B_{\underline{p}} - r \,.$$

Hence $(t_1,\dots,t_r)$ is part of a system of parameters of $B_{\underline{p}}$ by (ibid 16.5.6).

3.1.6. Finite morphisms of formal schemes. We recall that a morphism $f: \mathfrak{X} \to \mathcal{S}$ of (locally noetherian) schemes is called adic if for some Ideal of definition $\mathcal{T}$ of $\mathcal{S}$ the $f^*(\mathcal{T})\underline{O}_{\mathfrak{X}}$ is an Ideal of definition for $\mathfrak{X}$; every Ideal of definition on $\mathcal{S}$ has then the same property (EGA I 10.12.1). An adic morphism $f: \mathfrak{X} \to \mathcal{S}$ is called finite (EGA III 4.8.2) if the corresponding $f_0: X_0 = (\mathfrak{X}, \underline{O}_{\mathfrak{X}}/f^*(\mathcal{T})\underline{O}_{\mathfrak{X}}) \to S_0 = (\mathcal{S}, \underline{O}_{\mathcal{S}}/\mathcal{T})$ is finite. If $f: \mathfrak{X} \to \mathcal{S}$ is finite then $f_*(\underline{O}_{\mathfrak{X}})$ is a coherent $\underline{O}_{\mathcal{S}}$-Algebra (EGA III 4.8.6) and conversely such a coherent $\underline{O}_{\mathcal{S}}$-Algebra $\mathcal{A}$ determines a formal $\mathcal{S}$-scheme, finite over $\mathcal{S}$ (EGA III 4.8.7), which we denote by $\mathrm{Spf}(\mathcal{A})$; the so-called formal spectrum of $\mathcal{A}$ over $\mathcal{S}$. Moreover we have the following formula (which is a slight generalization of EGA III 4.8.8):

(*)
$$\mathrm{Hom}_{\mathcal{S}}(\mathfrak{X}', \mathrm{Spf}\,\mathcal{A}) \simeq \mathrm{Hom}_{\underline{O}_{\mathcal{S}}}(\mathcal{A}, g_*(\underline{O}_{\mathfrak{X}'})) \,,$$

where $g: \mathfrak{X}' \to \mathcal{S}$ is an arbitrary morphism. The right hand side means homomorphisms of $\underline{O}_{\mathcal{S}}$-Algebras (they are automatically continuous).

Note that the topology in $\mathcal{A}$ is determined by $\underline{O}_{\mathcal{S}}$, but the topology in $g_*(\underline{O}_{\mathcal{S}'})$ is not. The proof is the same as in EGA II 1.2.7 using EGA 0_I 7.6.6 instead of EGA 0_I1.2.4.

Lemma 3.1.7. Let $\mathcal{S}$ = Spf A, $\mathcal{S}'$ = Spf A' with A, A' adic noetherian rings and $f: \mathcal{S}' \to \mathcal{S}$ a morphism. Equivalent conditions:

i) f flat (resp. faithfully flat provided f is adic).

ii) Spec $A' \to$ Spec A flat (resp. faithfully flat provided f is adic).

Proof: For $s' \in \mathcal{S}'$ and $s = f(s')$ consider the following commutative diagram, where completion means J-adic (resp J'-adic) completion:

$$\begin{array}{ccc} \hat{\underline{O}}_{S,s} & \xrightarrow{\hat{\alpha}} & \hat{\underline{O}}_{S',s'} \\ \uparrow & & \uparrow \\ \underline{O}_{\mathcal{S},s} & \xrightarrow{\beta} & \underline{O}_{\mathcal{S}',s'} \\ \uparrow & & \uparrow \\ \underline{O}_{S,s} & \xrightarrow{\alpha} & \underline{O}_{S',s'} \end{array}$$

with faithfully flat vertical arrows. (cf. 3.1.2.)

i) $\Rightarrow$ ii) From the flatness of β we get the flatness of α. Hence we have flatness at every closed point $s' \in S'$, hence ([3], Alg. Comm. II,§ 3,prop.15.) A' is A-flat. Next take $s_o \in S$, then there is a closed point $s \in \overline{\{s_o\}}$ and a point $s' \in S'$ over s. From the flatness of $\underline{O}_{S,s} \to \underline{O}_{S',s'}$ follows the existence of a point $s'_o \in S'$ over s_o. Hence Spec $A' \to$ Spec A faithfully flat.(Note: f adic is not needed here.)

ii) $\Rightarrow$ i) From the flatness of α follows the flatness of $\hat{\alpha}$ ([3], Alg. Comm.III,§ 5 prop.4 and prop.2.), next the flatness of β. If $J' = J.A'$ then Spec $A' \to$ Spec A surjective implies $\mathcal{S}' - \mathcal{S}$ surjective.

3.1.8. $\mathcal{S}$-groups According to the general definitions (EGA 0_{III} 8.2) a formal $\mathcal{S}$-scheme $\mathcal{G}$ is a $\mathcal{S}$ -group if for every formal $\mathcal{S}$-scheme $\mathcal{T}$ the set

$$\mathcal{G}(\mathcal{T}) = \mathrm{Hom}_{\mathcal{S}}(\mathcal{T}, \mathcal{G})$$

is a group and for $\mathcal{T}_1 \to \mathcal{T}_2$ the corresponding map $\mathcal{G}(\mathcal{T}_2) \to \mathcal{G}(\mathcal{T}_1)$ a group homomorphism.

Remark: Note the difference between a $\mathcal{S}$-group and the notion of "formal group".

Examples: a) $\mathcal{G}$ an ordinary group, then $\mathcal{G}_{\mathcal{S}} = \bigoplus_{\mathcal{G}} \mathcal{S}$ is a $\mathcal{S}$-group; such a $\mathcal{S}$-group is called constant.

b) $\mu_{n,\mathcal{S}}$ and $\mu_{\underline{n},\mathcal{S}}$ defined similar as in 1.1.3.

3.1.9. Operation of a $\mathcal{S}$-group $\mathcal{G}$ on a formal $\mathcal{S}$-scheme $\mathcal{X}$. Again following EGA 0_{III} 8.2 we require for variable $\mathcal{S}$-scheme $\mathcal{T}$ a group operation of $\mathcal{G}(\mathcal{T})$ on $\mathcal{X}(\mathcal{T}) = \mathrm{Hom}_{\mathcal{S}}(\mathcal{T}, \mathcal{X})$, behaving functorially.

Example. For a set of regular sections $a_i \in \Gamma(\mathcal{S}, \underline{O}_{\mathcal{S}})$ ($i \in I$) and integers n_i prime to the residue characteristics define a formal $\mathcal{S}$-scheme $\mathcal{Z}_{\underline{n}}^{\underline{a}}$ similarly as in 1.2. In precisely the same way as in 1.2. we see that $\mu_{\underline{n}}$ operates on $\mathcal{Z}_{\underline{n}}^{\underline{a}}$ over $\mathcal{S}$; the couple $(\mathcal{Z}_{\underline{n}}^{\underline{a}}, \mu_{\underline{n}})$ is a Kummer covering of $\mathcal{S}$ relative to $\underline{a} = (a_i)_{i \in I}$. This formation is stable under base change (provided the sections remain regular) and also lemma 1.2.5 holds.

3.1.10. Quotient formation (in a very special case). Let $f: \mathcal{X} \to \mathcal{S}$ be a finite formal $\mathcal{S}$-scheme and $\mathcal{G}$ a constant group operating on $\mathcal{X}$ over $\mathcal{S}$.

Lemma There exists a quotient $\mathcal{X}/\mathcal{G} = \mathcal{Y}$, i.e., a formal $\mathcal{S}$-scheme $\mathcal{Y}$ and a $\mathcal{S}$-morphism

$$\varphi: \mathcal{X} \to \mathcal{Y}$$

such that:

i) as a topological space $\mathcal{Y}$ is the quotient of $\mathcal{X}$ by the group action,

ii) the structure sheaf is the sheaf of invariants

$$\underline{O}_{\mathcal{Y}} = f_*(\underline{O}_{\mathcal{X}})^{\mathcal{G}},$$

iii) $\varphi: \mathcal{X} \to \mathcal{Y}$ is the cokernel (in the category of affine formal $\mathcal{S}$-scheme which are adic over $\mathcal{S}$) of the couple

$$u,v: \mathcal{G} \times \mathfrak{X} \rightrightarrows \mathfrak{X}$$

with

$$u(g,x) = x, v(g,x) = g.x \quad .$$

Moreover $\mathcal{Y}$ is a formal $\mathcal{S}$-scheme and the formation of $\mathcal{Y}$ commutes with flat base change.

Proof: It suffices to prove this <u>locally on $\mathcal{S}$</u>, i.e., we can assume $\mathcal{S}$ = Spf A with A a J-adic noetherian ring and $\mathfrak{X}$ =Spf B. In this case take $\mathcal{Y}$ = Spf $B^{\mathcal{G}}$, then ii) and iii) (by EGA II 1.2.7 for formal adic $\mathcal{S}$-schemes) are satisfied, i) is satisfied by SGA1 V 1.1. (restricted to <u>open</u> primes). Furthermore the finiteness follows from A noetherian and B a finite A-module. The compatibility with flat base change follows from SGA 1 V 1.9.

3.2. Étale coverings of formal schemes

<u>3.2.1.</u> Let $f: \mathfrak{X} \to \mathcal{S}$ be a finite morphism (and as usual $\mathcal{S}$, hence $\mathfrak{X}$, a locally noetherian formal scheme). The coherent $\underline{O}_{\mathcal{S}}$-Algebra $f_*(\underline{O}_{\mathfrak{X}})$ (3.1.6) is denoted by $\mathcal{B}$. Furthermore $\mathcal{J}$ denotes an Ideal of definition for $\mathcal{S}$.

<u>Definition 3.2.2.</u>: $f: \mathfrak{X} \to \mathcal{S}$ is called an <u>étale covering</u> of $\mathcal{S}$ ("revetement étale") if:

1) f is finite,
2) $f_*(\underline{O}_{\mathfrak{X}})$ is locally free,
3) for every $s \in \mathcal{S}$ the (usual) scheme $f_s: \mathfrak{X}_s = \mathfrak{X} \times_{\mathcal{S}} k(s)$ is unramified over k(s).

<u>Remarks:</u> a) Compare the definition with EGA IV 18.2.3.

b) We have, if $S_o = (V(\mathcal{J}), \underline{O}_{\mathcal{S}}/\mathcal{J})$ and $X_o = \mathfrak{X} \times_{\mathcal{S}} S_o$, that $\mathfrak{X}_s = X_o \times_{S_o} k(s)$. Therefore 3 makes sense.

c) In section 6.1. we shall define the notion of <u>étale morphism</u> of étale schemes in general; we shall see that 3.2.2 is a special case of that notion.

Proposition 3.2.3. Let $f: \mathfrak{X} \to \mathcal{S}$. Equivalent conditions:

1) f is an étale covering.

2) for every $s \in \mathcal{S}$ the morphism

$$f_s^*: \operatorname{Spec} \mathcal{B}_s \to \operatorname{Spec} \underline{O}_{\mathcal{S},s}$$

is an étale covering of usual schemes (here $\mathcal{B}_s$ denotes the stalk of $\mathcal{B}$ at s).

3) If $\mathcal{S}$ = Spf A and $\mathfrak{X}$ = Spf B then the corresponding f: Spec B $\to$ Spec A is an étale covering.

Proof: First two remarks: If U= Spf A is an affine noetherian neighbourhood of s in $\mathcal{S}$ and B= $\Gamma(U, \mathcal{B})$ then we have

$$\mathcal{B}_s = \varinjlim_{f \in A;\ f(s) \neq 0} B \otimes_A A_{\{f\}} ,$$

where we can take the usual tensor product (EGA 0_I 7.7.8).

If f: X $\to$ S is a finite morphism of usual schemes then the set of points $s \in S$ for which there exists a point $x \in X$ with f(x)= s such that f is ramified at x, is a closed set (EGA IV 17.3.7).

1) $\Rightarrow$ 2) $\mathcal{B}_s$ is a finite and free $\underline{O}_{\mathcal{S},s}$-module, f_s^* is non ramified over s, hence nowhere by the above remark. The assertion follows from EGA IV 18.2.3.

2) $\Rightarrow$ 3) By [3],II, Th 1 of § 5, B is a projective A-module. The morphism is non-ramified over $V(\mathcal{J})$, hence nowhere. Apply EGAIV 18.2.3. It suffices now, since the assertions are local, to prove 3) $\Rightarrow$ 1).

3) $\Rightarrow$ 1) Clear because only condition 3) of 3.2.2 remains to be checked, but the closed fibres are the same for Spec A and Spf A.

3.2.4. Let RevEt($\mathcal{S}$) denote the full subcategory of formal $\mathcal{S}$-schemes $f: \mathfrak{X} \to \mathcal{S}$, for which f is an étale covering (cf.also 2.4.1). Put for every integer $n \geq 0$

$$S_n = (V(\mathcal{J}), \underline{O}_{\mathcal{S}} / \mathcal{J}^{n+1}) .$$

Theorem The natural functors

$$\mathfrak{X} \rightsquigarrow X_n = \mathfrak{X} \times_{\mathcal{S}} S_n \rightsquigarrow X_o = X_n \times_{S_n} S_o ,$$

from

$$\text{RevEt}(\mathcal{S}) \to \text{RevEt}(S_n) \to \text{RevEt}(S_o) ,$$

are equivalences.

Proof: Use EGA IV 18.1.2 and EGA I 10.12.3. Starting with a projective system $\{X_n\}$, the finiteness of $\mathfrak{X} = \varprojlim X_n$ follows from EGA I 10.11.3, the flatness from [3],III § 5 Th 1 and condition 3 of 3.2.2 is trivial.

Corollary 3.2.5. Let $\mathcal{S}$ be connected. Then RevEt($\mathcal{S}$) is a Galois category.

Proof: We have to make a fibre functor. Take $s \in \mathcal{S}$ and Ω a separably closed field containing k(s). Then we have a geometric point

$$\xi : \text{Spec}\,\Omega = \text{Spf}\,\Omega \to \mathcal{S} .$$

For $\mathfrak{X} \in$ RevEt($\mathcal{S}$) we have clearly $\mathfrak{X}_\xi \simeq (X_o)_\xi$ and the corollary follows from 3.2.4 and SGA 1 V 4. applied on S_o .

Corollary 3.2.6. There exists a profinite group $\pi_1(\mathcal{S}, \xi)$ such that the category RevEt($\mathcal{S}$) is equivalent with the category of finite sets on which this group operates continuously. The equivalence is obtained by means of the fibre functor

$$\mathfrak{X} \rightsquigarrow F(\mathfrak{X}) = \text{Hom}_{\mathcal{S}}(\text{Spf}\,\Omega , \mathfrak{X})$$

(where $F(\mathfrak{X})$ -by abuse of language- may be identified with $\mathfrak{X}_\xi$). The group $\pi_1(\mathcal{S}, \xi)$ is called the fundamental group of $\mathcal{S}$ with base point ξ. The "usual properties" hold by change of base point. Furthermore there are canonical isomorphisms

$$\pi_1(S_o, \xi) \xrightarrow{\sim} \pi_1(S_n, \xi) \xrightarrow{\sim} \pi_1(\mathcal{S}, \xi).$$

§4. Tamely ramified coverings of formal schemes

4.1. Definitions and elementary properties

4.1.1. Let $\mathcal{S}$ be a locally noetherian, normal formal scheme and D a closed subset on $\mathcal{S}$. Note that for every $s \in \mathcal{S}$ Spec $\underline{O}_{\mathcal{S},s}$ is a normal scheme. We say that D has codimension at least one if for every $s \in \mathcal{S}$ the corresponding closed subset D_s in Spec $\underline{O}_{\mathcal{S},s}$ (see 3.1.4) has codimension at least one. Note that it may happen that D is of codimension at least one and that supp(D)= $\mathcal{S}$! Let furthermore $f: \mathcal{X} \to \mathcal{S}$ be a finite morphism and write $f_*(\underline{O}_{\mathcal{X}}) = \mathcal{B}$; by $\mathcal{B}_s$ we denote the stalk at s; clearly $\mathcal{B}_s$ is a finite $\underline{O}_{\mathcal{S},s}$-algebra (cf.3.2.3).

4.1.2. Let D be a closed subset of codimension at least one on $\mathcal{S}$.

Definition A finite morphism $f: \mathcal{X} = \text{Spf}\, \mathcal{B} \to \mathcal{S}$ is called a tamely ramified covering of $\mathcal{S}$ relative to D if for every $s \in \mathcal{S}$ we have that

$$\text{Spec}\, \mathcal{B}_s \to \text{Spec}\, \underline{O}_{\mathcal{S},s}$$

is a tamely ramified covering of Spec $\underline{O}_{\mathcal{S},s}$ relative to D_s.

Remarks: 1) By abuse of language we often call $\mathcal{X}$ itself a tamely ramified covering of $\mathcal{S}$ relative to D, or a covering of $\mathcal{S}$ tame over D. We say shortly: $\mathcal{X}$ tame over $\mathcal{S}$ (relative to D).

2) For $s \notin$ supp D we have by definition that Spec $\mathcal{B}_s \to$ Spec $\underline{O}_{\mathcal{S},s}$ is an étale covering.

Lemma 4.1.3. Let $\mathcal{S}$ = Spf A, with A a J-adic ring (always noetherian). Suppose $\mathcal{S}$ is normal; put S= Spec A. Let $(D_i)_{i \in I}$ be a set of regular divisors with normal crossings on $\mathcal{S}$; denote the corresponding divisors on S by the same letter (see 3.1.5 ; note that they have the same property). Put $D = \sum_{i \in I} D_i$ (both on $\mathcal{S}$ and S). Finally let $\mathcal{X}$ = Spf B and X= Spec B with B a finite A-algebra (with the JB-adic topology).

Then the following conditions are equivalent:

i) X tame over S relative to D,

ii) $\mathfrak{X}$ tame over $\mathcal{S}$ relative to D.

Proof: First note that S is normal (3.1.3). Let $s \in \mathcal{S}$; consider the stalks

$$B_s = B \otimes_A \underline{O}_{S,s} \text{ and } \mathcal{B}_s = B \otimes_A \underline{O}_{\mathcal{S},s}\ .$$

We have a cartesian diagram

$$\begin{array}{ccc} \operatorname{Spec} B_s & \longleftarrow & \operatorname{Spec} \mathcal{B}_s \\ \downarrow \lambda & & \downarrow \mu \\ \operatorname{Spec} \underline{O}_{S,s} & \longleftarrow & \operatorname{Spec} \underline{O}_{\mathcal{S},s} \end{array}$$

i) $\Longrightarrow$ ii) The assumptions of 2.3.6 are fulfilled (cf.3.1.5); therefore i) $\Longrightarrow$ ii) by 2.3.6.

ii) $\Longrightarrow$ i) First consider $s \in \mathcal{S}$. The tame ramification of λ follows, by 2.2.9 , from the tame ramification of μ and the faithfully flatness of the lower horizontal arrow. Next take $s \in S$ arbitrary (i.e., not necessarily closed). There exists a point $s_1 \in \mathcal{S}$ which is a specialization of s and we have a cartesian diagram

$$\begin{array}{ccc} \operatorname{Spec} B_{s_1} & \longleftarrow & \operatorname{Spec} B_s \\ \downarrow & & \downarrow \\ \operatorname{Spec} \underline{O}_{S,s_1} & \longleftarrow & \operatorname{Spec} \underline{O}_{S,s} \end{array}$$

Take the maximal points of the divisors D_i in both schemes Spec $\underline{O}_{S,s}$ and apply the last remark of 2.2.8. This completes the proof.

Corollary 4.1.4. Let S be a normal scheme, $Y \subset S$ a closed subscheme and $(D_i)_{i \in I}$ regular divisors with normal crossings on S; put $D = \sum_{i \in I} D_i$. Consider the completion $\hat{S} = S_{/V(Y)}$ of S along Y (EGA I 10.8.5); suppose that $\hat{S}$ is normal. Let $j: \hat{S} \to S$ be the canonical morphism of ringed spaces. Then $(j^*(D_i)_{i \in I}$ is a set of regular divisors with

<u>normal crossings</u> on $\hat{S}$ and $j^*(D) = \sum_i j^*(D_i)$. Finally let $f: X \to S$ be tame over S relative to D, then $\hat{f}: \hat{X} = X_{/V(Y)} \to \hat{S}$ is tame over $\hat{S}$ relative to $j^*(D)$.

<u>Proof:</u> It suffices to prove this in case S= Spec A with A a noetherian ring and Y defined by an ideal J. Then $\hat{S}$= Spf $\hat{A}$ with $\hat{A}$ the J-adic completion of A. Introduce also $S' =$ Spec $\hat{A}$ and $X' = X \times_A \hat{A}$. By the remarks made in 3.1.5 and by 4.1.3 it suffices to prove that the inverse images $(D_i')_{i\in I}$ on S', of the $(D_i)_{i\in I}$ on S, are regular and have normal crossings and that X' is tame over S' (note that the assumptions on $\hat{S}$ imply that S' is normal, see 3.1.3). In order to see that the regularity and the normal crossings on S imply the same property on S' we can, by lemma 3.1.5[1], restrict to closed points $s' \in S'$; let s be the image of s' on S. If $\underline{m}$ (resp. $\underline{m}'$) is the ideal corresponding with s (resp. s')then we have by [3] III, §3 prop.8 that $\underline{m}' = \underline{m}.\hat{A}$, that $A_{\underline{m}} \subset \hat{A}_{\underline{m}'}$ and that both local rings have isomorphic completions (this time completion means completion with respect to the maximal ideals!). From this follows (going via the completions) that $A_{\underline{m}}$ regular implies $\hat{A}_{\underline{m}'}$ regular and that a regular system of parameters in $A_{\underline{m}}$ gives a regular system of parameters in $\hat{A}_{\underline{m}'}$. This completes the proof as far as the divisor is concerned. Now the assumptions of 2.3.6 are fulfilled for the morphism Spec $A' \to$ Spec A and this completes the proof.

<u>Corollary 4.1.5.</u> The same assumptions on S, Y, $\hat{S}$ and the divisors $(D_i)_{i\in I}$ on S. Let $f: X \to S$ be <u>finite</u>, with X <u>normal.</u> Suppose $\hat{f}: \hat{X} = X_{/V(Y)} \to \hat{S}$ tamely ramified over $\hat{S}$ relative to the divisor $j^*(D)$. Then there exists a Zariski open neighbourhood V of Y on S such that

$$f|V : X|V \to V$$

is tamely ramified relative to $D|V$.

Proof: Again we can assume $S = \operatorname{Spec} A$, $\hat{S} = \operatorname{Spf} \hat{A}$ and put $S' = \operatorname{Spec} \hat{A}$. Consider the cartesian diagram (with $s' \in S'$ over the point $s \in S$)

$$\begin{array}{ccc} X \times_A A_s & \longleftarrow & X \times_A (\hat{A})_{s'} \\ \downarrow{\scriptstyle \lambda} & & \downarrow{\scriptstyle \mu} \\ \operatorname{Spec} A_s & \stackrel{\alpha}{\longleftarrow} & \operatorname{Spec} (\hat{A})_{s'} \end{array}$$

From 4.1.3 we have that μ is tame. Since α is faithfully flat we can apply 2.2.9 , hence λ is tame. The corollary follows then from 2.2.10.

4.2. The category $\operatorname{Rev}^D(\mathcal{S})$

4.2.1. Let $\mathcal{S}$ be a locally noetherian, normal and connected formal scheme and $(D_i)_{i \in I}$ a locally finite set of regular divisors with normal crossings on $\mathcal{S}$. Put $D = \sum_{i \in I} D_i$. Consider in the category of formal $\mathcal{S}$-schemes the following full subcategories:

$\operatorname{Rev}(\mathcal{S})$: the formal $\mathcal{S}$-schemes which are finite over $\mathcal{S}$,

$\operatorname{RevEt}(\mathcal{S})$: the formal $\mathcal{S}$-schemes which are étale coverings of $\mathcal{S}$,

$\operatorname{Rev}^D(\mathcal{S})$: the formal $\mathcal{S}$-schemes which are tamely ramified over $\mathcal{S}$ relative to D.

We have the following inclusions:

$$\operatorname{RevEt}(\mathcal{S}) \to \operatorname{Rev}^D(\mathcal{S}) \to \operatorname{Rev}(\mathcal{S}).$$

Proposition 4.2.2. $\operatorname{Rev}^D(\mathcal{S})$ is a Galois category (after construction of a suitable fibre functor; see below).

Proof: We check the conditions of SGA 1 V 4. The conditions G_2, G_3 follow immediately from 4.1.3 , from the corresponding assertions for usual schemes, from the way the quotient is constructed (3.1.10) and from the remark that the splitting of a morphism into a strict epimorphism and a monomorphism can be preformed locally because it is unique.

As to G_1, i.e., the existence of fibre products, this is more subtle.

Let $f:\mathfrak{X} \longrightarrow \mathfrak{Z}$, $g:\mathfrak{Y} \rightarrow \mathfrak{Z}$, $h:\mathfrak{Z} \rightarrow \mathcal{S}$ be morphisms with h, h.f and h.g in $\mathrm{Rev}^D(\mathcal{S})$. <u>Assume first</u> that $\mathcal{S}$ = Spf A with A a noetherian J-adic ring; let $\mathfrak{Z}$ =Spf D, $\mathfrak{X}$ = Spf B, $\mathfrak{Y}$ = Spf C. Put S= Spec A, Z= Spec D, etc. Consider $(X \times_Z Y)_n$, where the subscript n means the <u>normalization</u> of S in the function ring of $X \times_Z Y$. Finally let

$$(X \times_Z Y)^{\sim}$$

denote the completion of $(X \times_Z Y)_n$ along V(J) on S (EGA I 10.8.5). By the construction of the fibre product in $\mathrm{Rev}^D(S)$ (see proof 2.4.2) and by 4.1.3 we have that $(X \times_Z Y)^{\sim}$ is in $\mathrm{Rev}^D(\mathcal{S})$, moreover it is the fibre product in that category because $(X \times_Z Y)_n$ is the fibre product in $\mathrm{Rev}^D(S)$.

For the general case we must show that this construction is independent of the "choice of the open piece" Spf A$\subset \mathcal{S}$. Let Spf A_2 $\subset$ Spf $A_1 \subset \mathcal{S}$, write S_1= Spec A_1, S_2= Spec A_2, X_1,X_2 etc. From the remarks in 3.1.5 we see that the base change

$$\mathrm{Spec}\ A_2 \rightarrow \mathrm{Spec}\ A_1$$

fulfills the condition of 2.3.6. Therefore $(X_1 \times_{Z_1} Y_1)_n \times_{S_1} S_2$ is <u>normal</u>. Therefore this scheme is canonically isomorphic with $(X_2 \times_{Z_2} Y_2)_n$. The fact that

$$(X_2 \times_{Z_2} Y_2)^{\sim} \overset{\sim}{\rightarrow} (X_1 \times_{Z_1} Y_1)^{\sim} \times_{\mathrm{Spf}\ A_1} \mathrm{Spf}\ A_2$$

follows then from EGA I 10.9.7.

<u>4.2.3. Construction of a fibre functor</u>. Let $s \in \mathcal{S}$, take an affine neighbourhood V= Spf A of s in $\mathcal{S}$. Put V^*= Spec A and let s^* be a point of V^* outside all the divisors D_i^* on V^*, corresponding with the divisors $D_i|V$ on V and such that $s \in \overline{\{s^*\}}$. In case s is outside all D_i we take s^*= s. Let Ω be a separably closed field containing $k(s^*)$, then we have a geometric point

$$\xi^* : \mathrm{Spec}\,\Omega \rightarrow V^*$$

with centre s^*. For $\mathfrak{X} \in \mathrm{Rev}^D(\mathcal{S})$, let $\mathfrak{X}|V$= Spf B and put X^*= Spec B, then we have, by 4.1.3 , $X^* \in \mathrm{Rev}^D(V^*)$. Define the "fibre functor"

$$F: \mathrm{Rev}^D(\mathcal{S}) \to (\text{Finite sets})$$

by

$$F(\mathfrak{X}) = X^*_{\xi^*} = \mathrm{Hom}_{V^*}(\mathrm{Spec}\,\Omega, X^*) \qquad (\forall\, \mathfrak{X} \in \mathrm{Rev}^D(\mathcal{S})).$$

By going from $\mathfrak{X}$ to X^* we see immediately that the conditions G_4 and G_5 of SGA 1 V 4 are fulfilled. As to G_6, i.e., $F(u)\colon F(\mathfrak{X}) \to F(\mathfrak{Y})$ an isomorphism implies u an isomorphism, we proceed as follows: First F(u) an isomorphism implies that the corresponding morphism of the usual schemes over V is an isomorphism (2.4.2), and hence $u|V$ is an isomorphism. From the connectedness of $\mathcal{S}$ follows then that all fibre functors are isomorphic and hence u is an isomorphism.

Corollary 4.2.4. There exists a profinite group $\pi_1^D(\mathcal{S}, \xi^*)$ such that the category $\mathrm{Rev}^D(\mathcal{S})$ is equivalent with the category of finite sets on which this group operates continuously. The equivalence is obtained by means of a fibre functor F as described in 4.2.3. This group is called the tame fundamental group of $\mathcal{S}$ with respect to D and with base point ξ^*.

Remarks 4.2.5. We have similar properties as in 2.4.5 for change of the base point, base change $\mathcal{S}' \to \mathcal{S}$ (provided the inverse images of the divisors are defined and are regular divisors with normal crossings, see 2.3.6) and Galois objects. Furthermore if $\mathfrak{X} \in \mathrm{Rev}^D(\mathcal{S})$, then the connected components are also in $\mathrm{Rev}^D(\mathcal{S})$. This follows from EGA III 4.8.3 and 2.2.5 (which holds also for formal schemes as is seen by 4.1.3). Therefore

$$\mathfrak{X} \text{ connected in } \mathrm{Rev}^D(\mathcal{S}) \Leftrightarrow \mathfrak{X} \text{ connected as formal scheme.}$$

Since the same is true for $\mathrm{RevEt}(\mathcal{S})$ we have that the natural continuous homomorphism

$$\pi_1^D(\mathcal{S}, \xi^*) \to \pi_1(\mathcal{S}, \xi)$$

is surjective (ξ is a geometric point for $\mathcal{S}$).

4.3. Relation between the tamely ramified coverings of a formal scheme and those of a subscheme defined by an Ideal of definition

4.3.1. Let again $\mathcal{S}$ be a locally noetherian, normal formal scheme and $(D_i)_{i\in I}$ a finite set of regular divisors with normal crossings on $\mathcal{S}$; put $D=\sum_{i\in I} D_i$. Assume $\mathcal{J}$ is an Ideal of definition with the following properties:

1) $(\mathcal{S}, \mathcal{O}_{\mathcal{S}}/\mathcal{J}) = S_o$ is a normal scheme,

2) the inverse images $D_{i,o}$ of the divisors D_i on S_o exist and the $(D_{i,o})_{i\in I}$ are regular divisors with normal crossings on S_o. Put $D_o = \sum_{i\in I} D_{i,o}$.

Theorem 4.3.2. With the above assumptions, the functor

$$\phi : \mathrm{Rev}^{D}(\mathcal{S}) \to \mathrm{Rev}^{D_o}(S_o) ,$$

defined by

$$\phi(\mathfrak{X}) = \mathfrak{X} \times_{\mathcal{S}} S_o = X_o \qquad (\text{with } \mathfrak{X} \in \mathrm{Rev}^{D}(\mathcal{S})),$$

is an equivalence of categories.

4.3.3. Proof: Reduction to usual schemes. First of all the fact that $\phi(\mathfrak{X})$ is tamely ramified over S_o relative to the D_o follows from 4.1.3 (applied on an affine neighbourhood of $s\in\mathcal{S}$) and 2.3.6 (note that the conditions of 2.3.6 are fulfilled!). First we prove the faithfulness of ϕ; it suffices to do this locally. Next it suffices to prove the fully faithfulness locally, and finally the equivalence locally. Therefore we can make the following additional assumptions:

a) $\mathcal{S}$ = Spf A with A a J-adic noetherian ring, J corresponds with $\mathcal{J}$. Write S= Spec A.

b) The divisors are defined by global sections $(a_i)_{i\in I}$ on $\mathcal{S}$.

c) S_o= Spec A/J is connected, hence (using normality) irreducible (and then $\mathcal{S}$ and S are also irreducible).

As to notations: we denote the divisors on S corresponding with D_i on $\mathcal{S}$ by the same letter; the $(D_i)_{i\in I}$ on S are regular and have normal crossings (3.1.5[1]). For $\mathfrak{X} \in \mathrm{Rev}^D(\mathcal{S})$ we have $\mathfrak{X}$ = Spf B with B a topological A-algebra; write X= Spec B. Then by 4.1.3 we have an <u>equivalence</u> between $\mathrm{Rev}^D(\mathcal{S})$ and $\mathrm{Rev}^D(S)$, <u>therefore we can replace $\mathcal{S}$ by the usual scheme S= Spec A</u> (A a J-adic ring).

<u>4.3.4. Some remarks proceeding the proof of 4.3.2.</u>

a) Given $X \in \mathrm{Rev}^D(S)$, <u>we claim that there exists a Kummer covering</u> $Z_{\underline{n}}^{\underline{a}} = S'$ of S, relative to the set of divisors $D_i = \mathrm{div}(a_i)$, such that the <u>normalization</u> $(X \times_S S')_n = X^*$ of $X \times_S S'$ is <u>étale over</u> S'.

$$\begin{array}{ccccc} X & \xleftarrow{\varphi'} & X \times_S S' & \longleftarrow & (X \times_S S')_n = X^* \\ \downarrow & & \downarrow & & \\ S & \xleftarrow{\varphi} & S' & & \end{array}$$

In order to see this we can assume (for simplicity) that X is irreducible. Let L be a smallest Galois extension of the function field R(X) over R(S). The divisors D_i determine valuation rings in R(S) and inertia groups. Take integers n_i which are multiples of the orders of these inertia groups and take $S' = Z_{\underline{n}}^{\underline{a}}$ with $\underline{n} = (n_i)_{i \in I}$. Then X^* is étale over the maximal points of the $\bar{\varphi}^{1}(D)$ (Abhyankar's lemma, SGA 1 X 3.6). Next take $s' \in \bar{\varphi}^{1}(D)$; Spec $\underline{O}_{S',s'}$ is a regular scheme (1.8.5). Apply the purity theorem to X^* over Spec $\underline{O}_{S',s'}$; we have X^* étale over every $s' \in \bar{\varphi}^{1}(D)$, hence over S'.

b) <u>Action of $\mu_{\underline{n}}$</u> . From the action of $\mu_{\underline{n}}$ on S' we get actions on $X_{S'}$ and X^* compatible with the action on S'. We have (1.3.2) that $X_{S'} / \mu_{\underline{n}} = X$. Furthermore we get a S-morphism

$$X^*/\mu_{\underline{n}} \rightarrow X_{S'}/\mu_{\underline{n}} = X,$$

which is an isomorphism over $U = S - \bigcup_i \mathrm{supp}(D_i)$ and then, since both schemes are normal (because quotients of normal schemes are normal), we have actually an isomorphism. Hence $X^*/\mu_{\underline{n}} \xrightarrow{\sim} X$.

c) <u>Reduction modulo $\mathcal{J}$</u>. From the assumption on the divisors D_i follows that $S'_o = S' \times_S S_o$ is a Kummer covering of S_o with respect to the $D_{i,o}$; put $U_o = U \cap S_o$. Since $(X^*)_o = X^* \times_{S'} S'_o$ is étale over S'_o (by base change), hence normal, and since its restriction over U_o is equal to the restriction of $(X \times_S S')\times_{S'} S'_o = X_o \times_{S_o} S'_o$ over U_o, we have canonically

$$(X_o)^* \simeq (X^*)_o ,$$

where $(X_o)^*$ is the <u>normalization</u> of $X_o \times_{S_o} S'_o$. In view of this canonical isomorphism we identify these two schemes and denote them by X^*_o.

<u>4.3.5. Faithfulness of ϕ</u>. Let be given two morphisms $g,h: X \to Y$ for X and Y in $\mathrm{Rev}^D(S)$ such that $g_o = h_o : X_o \to Y_o$. Take a suitable Kummer covering S' of S such that both X^* and Y^* are étale over S' (4.3.4a). Then we have, with obvious notations, $g^*_o = h^*_o$. By EGA IV 18.3.2 applied over the base S' we get $g^* = h^*$. Since $X^* | \bar{\varphi}^{1}(U) = X_{S'} | \bar{\varphi}^{1}(U)$, etc., we have by flat descent on φ' (see 4.3.4 a) that $g | U = h | U$. Finally, U is schematically dense in S because it is the complement of a divisor, hence $X | U$ is schematically dense in X (EGA IV 11.10.5). Therefore $g = h$.

<u>4.3.6. Full faithfulness of ϕ</u>. Given X and Y in $\mathrm{Rev}^D(S)$ and $g_o : X_o \to Y_o$. We take a suitable Kummer covering S' of S as above and use the notations from above. From g_o we obtain $g^*_o : X^*_o \to Y^*_o$. Again by EGA IV 18.3.2 , applied on S' , we obtain $g^* : X^* \to Y^*$ in such a way that base change $S'_o \to S'$ gives g^*_o. Since g^*_o is compatible with the action of $\mu_{\underline{n},S_o}$, the same is true for g^* with respect to the action of $\mu_{\underline{n},S}$. This follows from the faithfulness of ϕ applied to the following diagram in $\mathrm{Rev}^D(S)$ <u>over S</u>:

$$\begin{array}{ccc} X^* & \xleftarrow{\text{operation}} & X^* \times \mu_{n,S} \\ \downarrow g^* & & \downarrow g^* \times \mathrm{id} \\ Y^* & \xleftarrow{\text{operation}} & Y^* \times \mu_{\underline{n},S} \end{array}$$

By quotient formation (see 4.3.4 b) we find a morphism $\tilde{g} : X \to Y$. It remains to be shown that $(\tilde{g})_o = g_o$. Since $X^* | \bar{\varphi}^{1}(U) = X_{S'} | \bar{\varphi}^{1}(U)$, etc., we have

$$\tilde{g}_{S'} | \bar{\varphi}^{1}(U) = g^* | \bar{\varphi}^{1}(U).$$

Hence

$$(\tilde{g}_o)_{S'_o} \Big|\, \bar{\varphi}^{-1}(U_o) = g_o^* \Big|\, \bar{\varphi}^{-1}(U_o) = (g_o)_{S'_o} \Big|\, \bar{\varphi}^{-1}(U_o) .$$

Using the fact that $X_o^* \big| \bar{\varphi}^{-1}(U_o)$ is schematically dense in $(X_o)_{S'_o}$ we have $(\tilde{g}_o)_{S'_o} = (g_o)_{S'_o}$ and finally, by flat descent $\tilde{g}_o = g_o$.

<u>4.3.7. Φ is an equivalence</u>. Let $X_o \in \mathrm{Rev}^{D_o}(S_o)$ be given. For simplicity we can assume that X_o is irreducible. We are looking for $\tilde{X} \in \mathrm{Rev}^D(S)$ such that $\tilde{X}_o \simeq X_o$. As in 4.3.4 a we can find a Kummer covering S'_o of S_o, relative to the divisors $D_{i,o}$, such that the normalization $(X_o \times_{S_o} S'_o)_n = X_o^*$ of $X_o \times_{S_o} S'_o$ is <u>étale</u> over S'_o. Furthermore we can assume that S'_o is obtained by base change from a Kummer covering S' of S. By EGA IV 18.3.2 , applied over S', there exists an étale covering X^* of S' such that $(X^*)_o \simeq X_o^*$. Furthermore by 2.2.5 we have that $X^* \in \mathrm{Rev}^D(S)$. Next, consider the group action

$$X_o^* \times \mu_{\underline{n},S_o} \to X_o^*$$

over S_o. From the fact that Φ is fully faithful we get that $\mu_{\underline{n},S}$ operates on X^* and operates in a way compatible with the action on S'. Take the quotient $\tilde{X} = X^*/\mu_{\underline{n}}$. This is a <u>normal</u> scheme and by 2.2.5 $\tilde{X} \in \mathrm{Rev}^D(S)$. It remains now to be seen that $\tilde{X}_o \simeq X_o$.

By the base extension we have a morphism ρ which makes the following diagram commutative

$$\begin{array}{ccc} & X_o^* & \\ \swarrow & & \searrow \\ X_o \simeq X_o^*/\mu_{\underline{n}} & \xrightarrow{\ \rho\ } & (X^*/\mu_{\underline{n}})_{S_o} = \tilde{X}_o \end{array}$$

Since all schemes under consideration are <u>normal</u>, it suffices to show that $\rho \big| (X_o | U_o)$ is an isomorphism.

In order to see this, remark first that there is a canonical morphism

$$X^* \big| \bar{\varphi}^{-1}(U) \to \tilde{X}_{S'} \big| \bar{\varphi}^{-1}(U)$$

and this is in fact an isomorphism, because these are étale coverings of U and it is easily seen, from the fact that the operation of $\mu_{\underline{n}}$

is compatible with the action on S', that we have an isomorphism on geometric fibres. By base change, and a similar remark for $X_o^*|\bar{\varphi}^1(U_o)$ we obtain that $\rho_{S'_o}|\bar{\varphi}^1(U_o)$ is an isomorphism. Hence $\rho|U_o$ (or better ρ restricted to $X_o|U_o$) is an isomorphism by flat descent $S'_o \longrightarrow S_o$. This completes the proof.

Corollary 4.3.8. The assumptions are the same as in 4.3.2. Assume moreover $\mathcal{S}$ (hence S_o) connected. Take a geometric point

$$\xi : \operatorname{Spec}\Omega \to S_o \to \mathcal{S}$$

outside the support of the divisors. Then we have canonically

$$\pi_1^{D_o}(S_o, \xi) \xrightarrow{\sim} \pi_1^D(\mathcal{S}, \xi).$$

4.4. Transitivity properties

4.4.1. First we state some results on reduced formal schemes and reduced formal subschemes.

Lemma. Let $\mathcal{S}$ be a locally noetherian formal scheme. For every $s \in \mathcal{S}$ put

$$\mathcal{N}_s = \operatorname{nilrad}(\underline{O}_{\mathcal{S},s}).$$

Then the $\mathcal{N}_s$ are the stalks of an Ideal $\mathcal{N}$. Furthermore:

i) For every quasi-compact open U in $\mathcal{S}$ (in particular: for every affine open U) we have

$$\Gamma(U,\mathcal{N}) = \operatorname{nilrad} \Gamma(U,\underline{O}_{\mathcal{S}}).$$

ii) $\mathcal{N}$ coherent $\Leftrightarrow$ $\exists$ formal closed subscheme $\mathcal{S}_{red}$ of $\mathcal{S}$ such that

a) $\mathcal{S}_{red}$ is reduced,

b) $\mathcal{S}_{red}$ is defined as subscheme by a nilpotent Ideal.

Moreover this $\mathcal{S}_{red}$ is unique and is defined by $\mathcal{N}$.

iii) If $\mathcal{N}$ is coherent then for $\operatorname{Spf} A_1 \subset \operatorname{Spf} A \subset \mathcal{S}$, N = nilrad A, N_1 = nilrad A_1, we have

$$N \otimes_A A_1 \xrightarrow{\sim} N_1 .$$

iv) If $\mathcal{S} = \bigcup_i \mathcal{S}_i$ ($\mathcal{S}_i$ open) then

$$\mathfrak{X}_{red} \text{ exists} \iff \mathfrak{X}_{i,red} \text{ exists for all } i .$$

v) If $\mathfrak{X} = \operatorname{Spf} A$ then

$$\mathfrak{X}_{red} \text{ exists} \iff (A_{red})_{\{f\}} \quad \text{is reduced} \quad (\forall f \in A).$$

Proof: For every open U put

$$\Gamma(U,\mathcal{N}) = \left\{ f \in \Gamma(U,\underline{O}_{\mathfrak{X}});\ f \in \mathcal{N}_s \text{ for } \forall s \in U \right\},$$

then

$$\varinjlim_{U,\, s\in U} \Gamma(U,\mathcal{N}) = \mathcal{N}_s \quad ,$$

therefore the $\mathcal{N}_s$ determine an Ideal $\mathcal{N}$ in $\underline{O}_{\mathfrak{X}}$.

i) It is clear that

$$\text{nilrad } \Gamma(U,\underline{O}_{\mathfrak{X}}) \subset \Gamma(U,\mathcal{N})$$

Furthermore every $f \in \Gamma(U,\mathcal{N})$ is locally nilpotent, therefore we have equality for quasi-compact U.

ii) $\Longrightarrow$ If $\mathcal{N}$ is coherent then it defines a closed subscheme (EGA 10.14.2) and it is clear that this subscheme has the required properties.

$\Longleftarrow$ Let $\mathfrak{X}_{red}$ be a closed subscheme defined by a coherent Ideal $\mathcal{N}'$ with properties a) and b), then we have by a) $\mathcal{N}'_s \supset \mathcal{N}_s$ and by b) $\mathcal{N}'_s \subset \mathcal{N}_s$, hence $\mathcal{N}' = \mathcal{N}$.

iii) Follows from ii) and i) and EGA I 10.10.8 and 10.10.2.

iv) Follows from ii) since coherence is a local property.

v) Let N (resp. N') be the nilradical of A (resp. $A_{\{f\}}$) . If $\mathfrak{X}_{red}$ exists then we have by iii)

$$N' = N \otimes_A A_{\{f\}} \quad (= N_{\{f\}}) \;,$$

hence

$$(A_{red})_{\{f\}} = A_{\{f\}} / N_{\{f\}} = A_{\{f\}} / N'$$

is reduced. Conversely if $(A_{red})_{\{f\}}$ is reduced then $N' \subset N_{\{f\}}$, but $N_{\{f\}} = N \otimes_A A_{\{f\}}$, hence contained in N' ; therefore for every $f \in A$ we have $N_{\{f\}}$ is the nilradical of $A_{\{f\}}$. But this means that $\mathcal{N} = N^{\Delta}$ (notation from EGA I 10.10), hence $\mathcal{N}$ is coherent.

Lemma 4.4.2. Let $\mathfrak{X}$ be a locally noetherian scheme, D a divisor on $\mathfrak{X}$ such that

a) $\mathfrak{X}$ is regular at the points of supp D ,

b) D_{red} exists and has local rings which are integral.

Then D_{red} is a divisor (i.e., this closed subscheme is locally defined by an Ideal of type $f.\underline{O}_{\mathfrak{X}}$, with f regular) and

$$D_{red} = \sum_{loc.finite} E_i$$

with E_i connected and integral divisors. Furthermore for suitable integers q_i (>o) we have

$$D = \sum_i q_i E_i \quad ,$$

Proof: In order to see that D_{red} is a divisor it suffices, since the Ideal $\mathcal{J}(D_{red})$ defining D_{red} is coherent (4.4.1), that $\mathcal{J}(D_{red})_x = (f_x).\underline{O}_{\mathfrak{X},x}$ for all $x \in supp D$, with f_x regular in $\underline{O}_{\mathfrak{X},x}$. Since the local ring $\underline{O}_{\mathfrak{X},x}$ is regular we have for the corresponding inverse image D_x in Spec $\underline{O}_{\mathfrak{X},x}$, of the divisor D, that $D_x = div(\varphi_1^{q_1} .. \varphi_s^{q_s})$ with φ_α irreducible regular elements in $\underline{O}_{\mathfrak{X},x}$. By 4.4.1 ii) we have $(D_{red})_x = div(\varphi_1 ... \varphi_s)$. This shows that D_{red} is a divisor. Next using the fact that D_{red} is integral we see that we have in fact $D_x = div(\varphi^q)$ and $(D_{red})_x = div(\varphi)$ with φ a section of $\underline{O}_{\mathfrak{X}}$ in a neighbourhood of x and irreducible in $\underline{O}_{\mathfrak{X},x}$. Due to the coherence we have $D = div(\varphi^q)$ and $D_{red} = div(\varphi)$ in a neighbourhood U of x and using the fact that D_{red} is integral we have that φ remains irreducible in all stalks $\underline{O}_{\mathfrak{X},y}$ with $y \in U \cap supp(D)$. Now if E_i are the connected components of D_{red} then we have (with a locally finite sum)

$$D_{red} = \sum_i E_i \quad ,$$

and clearly the E_i are integral divisors. In a neighbourhood U of $x \in supp(D)$ we have, as we have seen above, for a suitable i that $E_i = div(\varphi)$ and $D = q_i E_i$. Due to the connectedness of E_i we have that $q_i = q_i(U)$ remains constant along E_i, i.e., is independent of U and this gives globally the required expression.

4.4.3. Assumptions and notations. In the remaining part of 4.4 we assume that $\mathcal{Y}$ is a locally noetherian, normal formal scheme. D is a divisor on $\mathcal{Y}$ such that

$$D = \sum_{i \in I} D_i ,$$

with a locally finite sum and such that the $(D_i)_{i \in I}$ are regular divisors with normal crossings on $\mathcal{Y}$.

Furthermore we assume that we have a partition $I = I' \cup I''$, $I' \cap I'' = \emptyset$ such that

$$D_i \cap D_j = \emptyset \text{ for } i \neq j \text{ and } i \in I', j \in I' .$$

Write

$$D' = \sum_{i \in I'} D_i \text{ and } D'' = \sum_{i \in I''} D_i .$$

Proposition 4.4.4. The assumptions and notations are as in 4.4.3. Let $f: \mathfrak{X} \to \mathcal{Y}$ be in $\mathrm{Rev}^{D'}(\mathcal{Y})$. Then we have for every i:

a) the $f^{-1}(D_i)$ is a divisor on $\mathfrak{X}$ and $\mathfrak{X}$ is regular in every point of $f^{-1}(D_i)$ (note: by $f^{-1}(D_i)$ we mean the closed subscheme defined by $\mathcal{J}(D_i)\underline{O}_{\mathfrak{X}}$),

b) the $\{f^{-1}(D_i)\}_{red}$ exists and has integral local rings,

c) furthermore

$$\{f^{-1}(D_i)\}_{red} = \sum_j E_{ij}$$

with $(E_{ij})_{ij}$ a, locally finite, set of regular divisors with normal crossings on $\mathfrak{X}$ and

$$E_{i_1 j_1} \cap E_{i_2 j_2} = \emptyset \text{ for } i_1 j_1 \neq i_2 j_2 \text{ if } i_1 \text{ and } i_2 \in I',$$

d) integers $q_{ij} > 0$ such that

$$f^{-1}(D_i) = \sum_j q_{ij} E_{ij}$$

with $q_{ij} = 1$ for $i \in I''$.

Proof: Outline: first we prove a) and b). Then we have by 4.4.2

the decompositions in c) and d). Next we prove the properties of the divisors E_{ij} and the integers q_{ij} in c) and d). In this way we see that it suffices to prove the assertions locally. Therefore we can assume that $\mathcal{S}$ = Spf A with A a J-adic noetherian ring and $D_i = \operatorname{div}(t_i)$ with $t_i \in A$. Then $\mathfrak{X}$ = Spf B, with B a certain finite A-algebra.

A) Put S= Spec A, X= Spec B and consider the divisors $D_i = \operatorname{div}(t_i)$ on S (we use the same letters as for the divisors on $\mathcal{S}$). We are first going to prove the (partially trivial) corresponding statements on S and X (note that $X \in \operatorname{Rev}^{D'}(S)$ by 4.1.3). In this case it suffices to prove the statements after an étale base change; therefore we can assume by 2.3.4 and the assumption $D_i \cap D_j = \emptyset$ $(i \neq j,\ i,j \in I')$ that X is a Kummer covering of S relative to D_i $(i \in I')$. We have

$$X = \operatorname{Spec} A[(\tau_i)_{i \in I}] \quad ,$$

with

$$\tau_i^{n_i} = t_i \qquad (i \in I,\ n_i = 1 \text{ if } i \in I''),$$

and integers n_i invertible on S. Then

$$f^{-1}(D_i) = \operatorname{div}(t_i) \text{ and } f^{-1}(D_i)_{red} = \operatorname{div}(\tau_i)$$

and those $\tau_i (i \in I)$ which are different from units form part of a regular system of parameters at $\underline{O}_{X,x}$ (cf. proof 1.8.6). Furthermore from $D_i \cap D_j = \emptyset$ for $i \neq j$ and $i \in I'$, $j \in I'$ follows

$$\operatorname{div}(\tau_i) \cap \operatorname{div}(\tau_j) = \emptyset \qquad (i \neq j,\ i \in I',\ j \in I').$$

This completes the proof in case of usual schemes.

B) Returning to the case of formal schemes, take $x \in \mathfrak{X}$ and let $s = f(x)$. There is a commutative diagram:

$$\begin{array}{ccc} \underline{O}_{X,x} & \longrightarrow & \underline{O}_{\mathfrak{X},x} \\ \uparrow \alpha & & \uparrow \beta \\ \underline{O}_{S,s} & \longrightarrow & \underline{O}_{\mathcal{S},s} \end{array}$$

Due to the assumptions on the D_i we have that for $x \in f^{-1}(D_i)$, for some $i \in I$, all four local rings are regular (2.3.4 , 1.8.6 and 3.1.3). Furthermore from the flatness of α (by 2.3.5) follows, by 3.1.7 , the flatness of β, hence $f^{-1}(D_i)$ is a divisor. This completes the proof of a).

Next if $x \notin f^{-1}(D_i)$ with $i \in I'$ then α is étale and a system of regular parameters in $\underline{O}_{S,s}$ gives such a system in $\underline{O}_{X,x}$. From this and 3.1.3 follow the assertions b), c) and d) easily in this case.

Let therefore $x \in f^{-1}(D_{i_o})$ with $i_o \in I'$ and $s = f(x)$; due to the assumptions $D_i \cap D_j = 0$ for $i \neq j$ in I', and since it suffices to prove the assertion locally on $\mathcal{S}$, we can assume $I' = \{i_o\}$. For simplicity of notations we can assume moreover that $s \in D_i$ for all $i \in I''$. For a <u>sufficiently small neighbourhood U of x on X</u> we have by part A) that on X

$$(*) \quad \begin{cases} f^{-1}(D_i)_{red} = \operatorname{div}(\tau_i) \\ \\ f^{-1}(D_i) = \operatorname{div}(t_i) = \operatorname{div}(\tau_i^{n_i}) \text{ and } n_i = 1 \text{ for } i \neq i_o, \text{ where for} \end{cases}$$

all $y \in U$ the $(\tau_i)_{i \in I}$ is part of a regular system of parameters in $\underline{O}_{X,y}$. According to 3.1.3 the $(\tau_i)_{i \in I}$ are also a regular system of parameters in $\underline{O}_{\mathfrak{X},y}$ for $y \in U \cap \mathfrak{X}$. Furthermore the second relation of $(*)$ holds also in $U \cap \mathfrak{X}$. On $\mathfrak{X}$ the $f^{-1}(D_i)_{red}$ is defined in $\underline{O}_{\mathfrak{X},y}$ by the root $\mathcal{K}_y$ of the ideal $(t_i)\underline{O}_{\mathfrak{X},y}$. Clearly

$$(\tau_i)\underline{O}_{\mathfrak{X},y} \subset \mathcal{K}_y ,$$

but on the other hand we have also that $\underline{O}_{\mathfrak{X},y} / (\tau_i)\underline{O}_{\mathfrak{X},y}$ is integral by EGA 0_{IV} 17.1.8 , hence

$$(\tau_i)\underline{O}_{\mathfrak{X},y} = \mathcal{K}_y .$$

This proves b), moreover it follows that the relations $(*)$ hold <u>both</u>, in $U \cap \mathfrak{X}$ on $\mathfrak{X}$ and in U on X, and this completes the proof.

<u>Proposition 4.4.5</u>. The assumptions are as in 4.4.3. Assume moreover that $\mathcal{S}$ is regular. Let $f: \mathfrak{X} \to \mathcal{S}$ be in $\mathrm{Rev}^{D'}(\mathcal{S})$ (hence clearly in $\mathrm{Rev}^{D}(\mathcal{S})$). Put $D_* = f^{-1}(D)_{red}$ (exists by 4.4.4!). Then there is a functor

$$\phi : \mathrm{Rev}^{D_*}(\mathfrak{X}) \to \mathrm{Rev}^{D}(\mathcal{S})$$

defined for $g: \mathcal{Y} \to \mathcal{X}$ in $\mathrm{Rev}^{D_*}(\mathcal{X})$ by $\phi(\mathcal{Y},g) = (\mathcal{Y}, f.g)$. Moreover ϕ is an equivalence between $\mathrm{Rev}^{D_*}(\mathcal{X})$ and the couples $(\mathcal{Y},g)$ with $\mathcal{Y} \in \mathrm{Rev}^{D}(\mathcal{S})$ and $g: \mathcal{Y} \to \mathcal{X}$.

Proof: By 4.4.4 the D_* fulfills the conditions of 4.1.3. Note in particular that $\mathcal{X}$ is regular. For the points $x \in \mathcal{X}$ with $f(x) \in \mathrm{supp}(D)$ this follows from 4.4.4 ; for the other points $x \in \mathcal{X}$ it follows from the étaleness of f, the fact that regularity is maintained by étale morphisms of usual schemes and by 3.1.3.

Let $(\mathcal{Y},g)$ be as above, i.e., in $\mathrm{Rev}^{D_*}(\mathcal{X})$. In order to see that $(\mathcal{Y}, f.g)$ is in $\mathrm{Rev}^{D}(\mathcal{S})$ we may, by 4.1.3 , reduce to usual schemes and apply 2.2.5.

In order to see that ϕ is an equivalence we must start with $\mathcal{Y} \in \mathrm{Rev}^{D}(\mathcal{S})$, $g: \mathcal{Y} \to \mathcal{X}$. In order to see that $(\mathcal{Y},g) \in \mathrm{Rev}^{D_*}(\mathcal{X})$ we again reduce to usual schemes by 4.1.3 and apply 2.2.5.

Corollary 4.4.6. Same assumptions as in 4.4.5 with $\mathcal{S}$ and $\mathcal{X}$ connected. Let η^* be a geometric point on $\mathcal{X}$ and $\xi^* = f(\eta^*)$ its image on $\mathcal{S}$ (cf.4.2.3) Then $\pi_1^{D_*}(\mathcal{X},\eta^*)$ is an open subgroup of $\pi_1^{D}(\mathcal{S},\xi^*)$. In case $\mathcal{X}$ is Galois in $\mathrm{Rev}^{D}(\mathcal{S})$ with group Γ (see 4.2.5) we have an exact sequence:

$$1 \to \pi_1^{D_*}(\mathcal{X},\eta^*) \to \pi_1^{D}(\mathcal{S},\xi^*) \to \Gamma \to 1 .$$

Proof: 4.4.5 and SGA 1 V 6.13.

----

§5. The tame fundamental group of a formal neighbourhood of an irreducible divisor

5.1. Preliminary investigation of the relation between the tame fundamental group of an irreducible divisor and the tame fundamental group of its formal neighbourhood

5.1.1. Throughout §5 and §7 we make the following assumptions:

a) $\mathcal{S}$ is a connected, locally noetherian, regular formal scheme,

b) $(D_i)_{i=0,1,\dots,r}$ is a set of regular divisors with normal crossings on $\mathcal{S}$,

c) the Ideal $\mathcal{J}(D_0)$ of the divisor D_0 is an Ideal of definition for $\mathcal{S}$.

Note that -due to the fact that D_0 is reduced- $\mathcal{J}(D_0)$ is a maximal Ideal of definition for $\mathcal{S}$. Write

$$S_0 = (\mathcal{S}, \mathcal{O}_{\mathcal{S}}/\mathcal{J}(D_0)).$$

$\mathcal{S}$, S_0 and D_0 have the same underlying topological space.

We use the following notations:

$$D = \sum_{i=0}^{r} D_i \quad ,$$

$$D' = \sum_{i=1}^{r} D_i \quad ,$$

$$D'_0 = \sum_{i=1}^{r} D_{i,0} \qquad (D_{i,0} \text{ inverse image of } D_i \text{ on } S_0).$$

In order to avoid misunderstanding, we emphasis that, although the above assumptions and notations are used tacitly throughout §5 and 7, they will be strengthened at the appropriate places (notably at 7.2.)

Lemma 5.1.2. S_0 is regular and the $(D_{i,0})_{i=1,\dots,r}$ are a set of regular divisors with normal crossings on S_0.

Proof: The first assertion follows from EGA 0_{IV}17.1.8 , the remaining part from EGA 0_{IV} 17.1.9.

5.1.3. Choice of the base point, on S_o and on $\mathcal{S}$, in case we are dealing with the divisor D'.

Since S_o is connected and regular, it is irreducible. For the base point $\xi_o \in S_o$ (2.4.3) we take a geometric point centered at the generic point s_o of S_o; i.e., let Ω_o be a separable closure of $k(s_o)$ then

$$\xi_o : \operatorname{Spec} \Omega_o \to S_o \to \mathcal{S}$$

is a geometric point both for S_o and for $\mathcal{S}$. According to 4.3.8 we have canonically

$$\pi_1^{D'_o}(S_o, \xi_o) \xrightarrow{\sim} \pi_1^{D'}(\mathcal{S}, \xi_o).$$

Furthermore we have by 2.4.6 c) a canonical surjective homomorphism

(*) $$\operatorname{Gal}(\Omega_o / k(s_o)) \to \pi_1^{D'_o}(S_o, \xi_o).$$

5.1.4. Choice of the base point on $\mathcal{S}$ in case we are dealing with the divisor D. The above base point ξ_o is all right in case we have the divisor D', however, for D we must take the base point outside $\mathcal{S}$. We proceed as in 4.2.3 : Take an affine open set Spf $A = V$ on $\mathcal{S}$ (A is a noetherian J-adic ring, J corresponds with $\mathcal{J}(D_o)$) such that V (and hence $V_o = V \cap S_o$) has no point in common with the D_i (resp. $D_{i,o}$) for $i > o$. The $V_o = \operatorname{Spec} A_o$ (with $A/J = A_o$) is irreducible, its generic point s_o is the point used in 5.1.3. Put $V^* = \operatorname{Spec} A$; V is regular, hence V^* is (3.1.3) regular and (being connected) it is also irreducible. Let s^* be the generic point and Ω^* the separable algebraic closure of $k(s^*)$, then Ω^* defines a geometric point $\xi^* : \operatorname{Spec} \Omega^* \to V^*$ (see 4.2.3.) Let Ω_1^* be the subfield of Ω^* which is the compositum of those finite extensions of $k(s^*)$ which are unramified (i.e., étale, EGA IV 18.10.1) over V^*. Since the couple $(V^*, V_o = V_o^*)$ is henselian (EGA IV 18.5.16) it follows that Ω_1^* determines an extension Ω' of $k(s_o)$ which is isomorphic with the compositum Ω'_o in Ω_o (where Ω_o is the field introduced in 5.1.3) of all finite extension of $k(s_o)$ unramified over $V_o = V_o^*$. Namely, starting with an étale and irreducible covering W of V^*, we have $W_o = W \times_{V^*} V_o$ étale and connected over V_o, hence normal, hence irreducible. Therefore a finite extension within Ω^* determines a finite extension of $k(s_o)$ unramified over V_o and every such extension can be obtained in this way by EGA IV 18.3.2. Now we choose once for all an isomorphism between Ω' and Ω'_o. In this way we "have fixed a path"

$$\pi_1^{D'}(\mathcal{S},\xi^*) \xrightarrow{\sim} \pi_1^{D'}(\mathcal{S},\xi_o),$$

and in the following we shall call -by abuse of language- this isomorphism canonical. Note also that the homomorphism(*) of 5.1.3 factors though $\mathrm{Gal}(\Omega'_o/k(s_o))$ because, since $V_o \cap D_{i,o} = \emptyset$, we have that for $X \in \mathrm{Rev}^{D'_o}(S_o)$ the restriction over V_o is étale over V_o.

5.1.5. Clearly $\mathcal{X} \in \mathrm{Rev}^{D'}(\mathcal{S})$ implies $\mathcal{X} \in \mathrm{Rev}^{D}(\mathcal{S})$, therefore we have canonically

$$\pi_1^{D}(\mathcal{S},\xi^*) \to \pi_1^{D'}(\mathcal{S},\xi^*).$$

and this continuous homomorphism is surjective because connectedness in both Galois categories corresponds with connectedness of the spaces (see 4.2.5.) Let K be the kernel, i.e., we have an exact sequence

$$1 \to K \to \pi_1^{D}(\mathcal{S},\xi^*) \to \pi_1^{D'}(\mathcal{S},\xi^*) \to 1 .$$

In this section we want to make a preliminary investigation of K.

5.1.6. The $\mathcal{S}$-group μ^t. For each positive integer n we have the $\mathcal{S}$-group (3.1.8) $\mu_{n,S}$ (or shortly μ_n). Furthermore we have a canonical transition homomorphism (cf.1.1.4):

$$\varphi_{nn'} : \mu_{n'} \to \mu_n \qquad (\text{with } n|n').$$

Put

$$\mu^t = \varprojlim_{n \text{ inv}} \mu_n$$

where the limit is taken over the integers n invertible on $\mathcal{S}$, i.e., the n prime to all residue characteristics of $\mathcal{S}$. Note that

$$\mu^t(\xi_o) = \varprojlim_{n \text{ inv.}} \mu_n(\xi_o)$$

is an ordinary profinite group (of certain roots of unity in $\overline{k(s_o)}$).

Proposition 5.1.7. The assumptions on $\mathcal{S}$ and D are as in 5.1.1. Let $\mathcal{X} \in \mathrm{Rev}^D(\mathcal{S})$ be a connected, pointed Galois covering with group Γ (see 4.2.5) and let

$$\varphi: \pi_1^D(\mathcal{S}, \xi^*) \to \Gamma$$

be the corresponding surjective continuous homomorphism (see remark below); put $J = \varphi(K)$ (the kernel K from 5.1.5). Let $\eta_o \in \mathcal{X}_{\xi_o}$ (geometric fibre of $\mathcal{X}$ over ξ_o) and Γ_{η_o} the inertia group of η_o (see SGA 1 V page 7; i.e., the stabilizer of η_o under the action of Γ). Then

i) $J = \Gamma_{\eta_o}$.

ii) For suitable n, invertible on $\mathcal{S}$, there is a canonical isomorphism

$$\mu_n(\xi_o) \overset{\sim}{\to} J .$$

iii) Using this isomorphism of ii) and the homomorphism (*) of 5.1.3 the operation of $\pi_1^{D'_o}(S_o, \xi_o) \overset{\sim}{\to} \pi_1^{D'}(\mathcal{S}', \xi^*)$ (see 4.3.8 and 5.1.4) on J, by inner automorphisms of Γ, corresponds with the action of $\mathrm{Gal}(\overline{k(s_o)}/k(s_o))$ on the roots of unity $\mu_n(\xi_o)$.

5.1.8. First some remarks:

a) The fact that $\mathcal{X} \in \mathrm{Rev}^D(\mathcal{S})$ is pointed means that there is fixed a point $\eta^* \in X^*_{\xi^*}$. This corresponds in the Galois category with a unique continuous homomorphism φ used in 5.1.7 (see SGA 1 V).

b) J does not depend on the pointing η^* because it is a normal subgroup of Γ.

c) It will follow from the isomorphism ii) that J is abelian, therefore we have indeed an action of $\pi_1^{D'}(\mathcal{S}', \xi^*)$ on J via inner automorphisms of Γ.

Lemma 5.1.9. Let $\mathcal{X} \in \mathrm{Rev}^D(\mathcal{S})$ be connected. Then $X_o = \mathcal{X} \times_{\mathcal{S}} S_o$ is irreducible.

Proof: $\mathcal{X}$ is connected, hence X_o is connected. It suffices to prove that $X_{o,red}$ is normal. This is a local assertion; we can assume

$\mathcal{Y}$ = Spf A with A an I-adic ring and the divisors D_i defined by global sections. Let $\mathcal{X}$ = Spf B and put S= Spec A, X= Spec B, then $X \in Rev^D(S)$ (see 4.1.3) and we are reduced to <u>usual</u> schemes.

If $S' \to S$ is an étale base change then it suffices to prove that $(X_{o,red})_{S'_o}$ is normal (EGA IV 6.5.4) and since

$$(X_{S'})_{o,red} = X_{S'_{o,red}} \xrightarrow{\sim} (X_{o,red})_{S'_o}$$

(a reduced scheme remains reduced by the étale base change $S'_o \to S_o$, EGA IV 6.5.3), we can replace S by S'. Therefore we are (by 2.3.4) reduced <u>to the case that X is a generalized Kummer covering</u>.

In that case we have S= Spec A (regular, noetherian), X= Spec B with

$$B = A[\underline{t}^{\underline{\alpha}}]$$

with $\underline{\alpha} \in N$ and N a subgroup of $\mathbb{Z}_{\underline{n}}$ (the notations are from 1.3.1). Let $D_i = \operatorname{div}(a_i)$, $a_i \in A$ $(i=o,\dots,r)$. Then S_o= Spec A_o and X_o= Spec B_o with $B_o = B/(a_o)B$. Then it is easily checked that

$$B_{o,red} = A_o[\underline{\tau}^{\underline{\alpha}'}]$$

with $\underline{\tau} = (\tau_i)_{i \neq o}$, $\underline{\alpha}' = (\alpha_1,\dots,\alpha_r)$ such that $(o,\alpha_1,\dots,\alpha_r) \in N$ and multiplication table

$$\underline{\tau}^{\underline{\alpha}'} \cdot \underline{\tau}^{\underline{\beta}'} = \prod_{i \neq o} \bar{a}_i^{e_i} \, \underline{\tau}^{\underline{\gamma}'} ,$$

with e_i and γ_i determined by

$$\alpha_i + \beta_i = e_i n_i + \gamma_i \qquad \text{with} \qquad e_i = \begin{cases} o & \text{if } \alpha_i+\beta_i < n_i \\ 1 & \text{if } \alpha_i+\beta_i \geq n_i \end{cases}$$

and $\bar{a}_i$ the image of a_i in A_o .

However, then we see (cf. 1.3.2) that Spec $B_{o,red}$ is a generalized Kummer covering of Spec A_o with respect to the divisors $D_{i,o}$ and by 5.1.2 and 1.8.5 this is a normal scheme.

<u>5.1.10. Proof of 5.1.7.</u>

a) Γ_{η_o} is a <u>normal subgroup</u> of Γ. In order to see this let V= Spf A be as in 5.1.4 (i.e., $V \cap D_i = \emptyset$, $i \neq o$); let $\mathcal{X}|V$= Spf B. The divisor

D_o defines a prime ideal $\underline{p}$ in A, $X_o|V$ is defined by $\underline{p}B$ and $X_{o,red}|V$ by the root $\sqrt{\underline{p}.B} = \underline{p}_1$ which, by 5.1.9, is a prime ideal in B. The group Γ operates on B but leaves $\underline{p}.B$, and hence $\underline{p}_1$, fixed (the fact that $\underline{p}_1$ is a prime ideal could - in usual ramification theory terminology- be reformulated as: Γ itself is the decomposition group). By definition Γ_{η_o} is the subgroup operating trivial upon $B/\underline{p}_1$, i.e.,

$$\Gamma_{\eta_o} = \{\sigma ; \sigma(b) \equiv b(\text{mod } \underline{p}_1), \forall b \in B\} .$$

Since for every $\tau \in \Gamma$ we have $\tau(\underline{p}_1) = \underline{p}_1$, it follows that $\tau \Gamma_{\eta_o} \tau^{-1} = \Gamma_{\eta_o}$ which shows that Γ_{η_o} is a normal subgroup of Γ (we see also that Γ_{η_o} is independent of η_o, i.e., all $\eta_o \in \mathfrak{X}_{\xi_o}$ have the same inertia group).

b) $J \supset \Gamma_{\eta_o}$. Proof: The pointed object $(\mathfrak{X}/J, \eta_o)$ of $\text{Rev}^D(\mathcal{J})$ is, by the definition of J, in the Galois category $\text{Rev}^{D'}(\mathcal{J})$. Working over $V^* =$ Spec A, with usual schemes, we have if $X^* =$ Spec B, that X^*/J is in $\text{Rev}^{D'}(V^*)$, i.e., X^*/J is étale over s_o. Look to the corresponding coverings over the valuation ring $\underline{O}_{V^*,s_o}$. There we can apply the usual ramification theory and it is well known that X^*/J étale over s_o implies $J \supset \Gamma_{\eta_o}$ (see for instance [6], Corps locaux, Chap 1 prop.22).

c) $J \subset \Gamma_{\eta_o}$. Proof: Consider $\mathfrak{X}/\Gamma_{\eta_o} \in \text{Rev}^D(\mathcal{J})$; we have to show that this is in fact in $\text{Rev}^{D'}(\mathcal{J})$. First take $X^*/\Gamma_{\eta_o} \in \text{Rev}^D(V^*)$, we want to see that this is étale over s_o (the generic point of S_o). Now Γ_{η_o} is also the stabilizer of the point η_o considered as point of $X^*_{\xi_o}$ (because $X^*_{\xi_o} = \mathfrak{X}_{\xi_o}$ with the same action of Γ). By SGA 1 V 2.2 we have X^*/Γ_{η_o} is étale over s_o. Clearly the same thing remains true if we replace V by another affine set. Let $s \in D_o$, $s \notin D_i (i \neq o)$. Let $\mathfrak{X} = \text{Spf}\, \mathcal{B}$ and consider Spec $\mathcal{B}_s$ over Spec $\underline{O}_{\mathcal{J},s}$; since this is étale over the generic point s_o of D_o we have by the purity theorem (SGA 1 X 3.2) that Spec $\mathcal{B}_s$ is étale over Spec $\underline{O}_{\mathcal{J},s}$. Hence $\mathfrak{X}/\Gamma_{\eta_o}$ is at worst ramified over D'.

d) By b) and c) the proof of 5.1.7 i) is complete. Next replace $A = \Gamma(V, \underline{O}_{\mathcal{J}})$ by the discrete valuation ring A_{s_o} (s_o the generic point of D_o). Note that $\pi_1^D(S_o, \xi_o)$ operates via $\Gamma' = \Gamma/J$. Both assertions ii) and iii) are well-known [8] and follow from the tameness except for the fact that the integer n in ii) is prime to all residue characteristics.

e) Let therefore $s \in D_o$ be arbitrary. We have to prove that the order of the inertia group Γ_{η_o} is prime with the characteristic of k(s). Take an affine neighbourhood W= Spf A of s on $\mathcal{S}$; let $\mathfrak{X}|W$= Spf B . Consider W^*= Spec A and X^*= Spec B. Γ operates on X^* and since X^* and $\mathfrak{X}$ have the same geometric fibre over ξ_o we can consider the same inertia group Γ_{η_o} for X^*.

Replacing W^* by an étale covering we do not change the inertia group by SGA 1 V 2.1. Using this remark we can assume, by 2.3.4, that X^* is a union of generalized Kummer coverings and even that X^* consists of one generalized Kummer covering (replace X^* by a connected component, if necessary). However, in that case we have that Γ is of type D(N) (cf. proof of 2.3.4) and D(N), hence Γ_{η_o}, has order prime to the characteristic of $k(s_o)$ by the definition of Kummer coverings.

Corollary 5.1.11. The assumptions are as in 5.1.1. Let K be defined by the exact sequence (5.1.5) (note also 5.1.3 and 5.1.4):

$$1 \to K \to \pi_1^D(\mathcal{S}, \xi^*) \to \pi_1^{D'_o}(S_o, \xi_o) \to 1 .$$

Then there is a canonical continuous, surjective homomorphism

$$j: \mu^t(\xi_o) \to K \to 1 ,$$

and the action of $\pi_1^{D'_o}(S_o, \xi_o)$ on K (via inner automorphisms of $\pi_1^D(\mathcal{S}, \xi^*)$) corresponds by j and the homomorphism (*) of 5.1.3, with the action of the Galois group $\mathrm{Gal}(\overline{k(s_o)}/k(s_o))$ on the roots of unity $\mu^t(\xi_o)$ in $k(s_o)$.

Proof: Consider the projective limit

$$\pi_1^D(\mathcal{S}, \xi^*) = \varprojlim \Gamma ,$$

where Γ runs through the group of connected pointed Galois coverings (see 5.1.7). The corollary follows from 5.1.7 by passing to the limit since all isomorphisms in 5.1.7 are canonical.

5.2. Preliminary investigation of the extension (of 5.1.11)

5.2.1. The assumptions and notations are as in 5.1.1. As to the notations, put

$$U_o = S_o - \bigcup_{i \neq o} \operatorname{supp}(D_i)$$

and

$$\pi_1^t = \pi_1^{D_o'}(S_o, \xi_o).$$

Consider the system

$$(X_\alpha, \xi_\alpha) \in \operatorname{Rev}^{D_o'}(S_o)$$

of connected, pointed Galois coverings of S_o, tamely ramified with respect to D_o' and with pointing $\xi_\alpha \in (X_\alpha)_{\xi_o}$ and group Γ_α. This system is partially ordered and filtered; the partial ordening is given by

$$(X_\alpha, \xi_\alpha) > (X_\beta, \xi_\beta)$$

iff there exists $f_{\beta\alpha} : X_\alpha \to X_\beta$ with $f_{\beta\alpha}(\xi_\alpha) = \xi_\beta$. With (X_α, ξ_α) corresponds uniquely a continuous surjective homomorphism

$$\pi_1^t \to \Gamma_\alpha$$

and

$$\pi_1^t = \varprojlim \Gamma_\alpha .$$

Note that, since X_α is normal and connected, it is irreducible. Put

$$U_\alpha = X_\alpha \big| U_o ,$$

then U_α is étale over U_o; write symbolically

$$(\vec{U}_o^t, \tilde{\xi}) = \varprojlim_\alpha (U_\alpha, \xi_\alpha) .$$

5.2.2. Consider in the étale topology of U_o sheaves of abelian groups $\underline{F}$ satisfying the following 3 properties:

i) $\underline{F}$ is a locally constant, constructible, torsion sheaf.

It is well-known (SGA 4 IX 2.2) that such a sheaf corresponds with an étale covering of U_o, which we denote by the same letter $\underline{F}$. Consider the normalization of S_o in the function ring $R(\underline{F})$ of $\underline{F}$; this normalization is denoted by $\underline{F}^n$. We make the following two additional assumptions:

ii) The normalization $\underline{F}^n$ is tamely ramified over S_o relative to the divisor D_o'.

iii) The degree of $\underline{F}$ over S_o, i.e., the number of points in the geometric fibres of $\underline{F}$, is prime to the residue characteristics of S_o.

A sheaf $\underline{F}$ on U_o with property i) corresponds uniquely with a finite abelian group F on which $\pi_1(U_o, \xi_o)$ operates continuously; $\underline{F}$ satisfies ii) iff this operation factors through π_1^t by means of the homomorphism of (2.4.6 c). Therefore the sheaves $\underline{F}$ with property i), ii) and iii) correspond in a one to one manner with the finite abelian groups F on which π_1^t operates continuously and which have order prime to the residue characteristics of S_o .

5.2.3. Let $\underline{F}= (\underline{F}_\gamma)$ be a projective system of sheaves $\underline{F}_\gamma$, each $\underline{F}_\gamma$ satisfying the conditions of 5.2.2. According to the last remark in 5.2.2 it amounts to the same thing to give such a system $(\underline{F}_\gamma)$ or to give a profinite group $F= \varprojlim F_\gamma$ on which π_1^t operates continuously and with each F_γ of order prime to the residue characteristics.

Morphisms of such projective systems are defined in the usual way and correspond uniquely with continuous π_1^t-homomorphisms of the profinite groups in question.

Let U be étale over U_o. Put

$$H^i(U,\underline{F})= \varprojlim_\gamma H^i(U,\underline{F}_\gamma)$$

and also

$$H^i(\tilde{U}_o^t,\underline{F}) = \varprojlim_{\gamma} \left\{ \varinjlim_{\alpha} H^i(U_\alpha,\underline{F}_\gamma) \right\} .$$

Proposition 5.2.4. With the above notations and assumptions we have for $\underline{F} = (\underline{F}_\gamma)$:

a) $H^1(\tilde{U}_o^t, \underline{F}) = o$.

b) There exists a pro-finite abelian group $\pi_2^{D_o'}(U_o, \xi_o)$ (shortly denoted by π_2^t in the following) which fulfills the requirements of 5.2.3 and such that there exists a functorial isomorphism

$$H^2(\tilde{U}_o,\underline{F}) = \mathrm{Hom}_{\mathrm{cont},\pi_1^t}(\pi_2^t,F)$$

where F is the π_1^t-group corresponding with $\underline{F} = (\underline{F}_\gamma)$ according to 5.2.3.

Proof: a) It suffices to prove this in the case $\underline{F} = \underline{F}_\gamma$ (i.e., in case of one sheaf). Let be given

$$v_\alpha \in H^1(U_\alpha,\underline{F}) .$$

Due to the assumptions i) and ii) of 5.2.2 we can assume -replacing α if necessary by a larger index- that $\underline{F}\big| U_\alpha$ is a constant sheaf. Let $\underline{F}$ correspond with the finite abelian π_1^t- group F, then the open subgroup $\pi_1(U_\alpha, \xi_\alpha) = \pi^*$ of $\pi_1(U_o, \xi_o)$ operates trivial on F. The element $v_\alpha \in H^1(U_\alpha, F_{U_\alpha})$ determines a F-torsor over U_α, let Y_α be the normalization of S_o in the function ring of this torsor. This is a covering of X_α and the assertion a) will be proved if we can show that Y_α is tamely ramified over S_o with respect to D_o' because then we take $X_\beta = Y_\alpha$; clearly by a suitable choice of the pointing we have then $(X_\beta, —) > (X_\alpha, \xi_\alpha)$ and the image of v_α in $H^1(U_\beta, F_{U_\beta})$ is trivial, with $U_\beta = Y_\alpha \big| U$. First we need:

Lemma 5.2.5. Let S be a connected scheme, T a $\mathfrak{g}$-torsor in the étale topology of S with $\mathfrak{g}$ an ordinary finite group (i.e.,$\pi_1(S)$ operates trivial on it). Then $T = \coprod_{\rho} T_\rho$, with T_ρ connected Galois coverings of S and if deg (T_ρ) denotes the degree of T_ρ over S (i.e., the number of points in the geometric fibres) then the deg (T_ρ) divides the order $(\mathfrak{g})$.

Proof: In the Galois category Rev Et(S) the T corresponds with a finite set E, E is a $\mathfrak{g}$ -torsor and the operation of $\pi_1(S)$ and $\mathfrak{g}$ commute. Then

$$E= \bigcup E_\rho$$

with the E_ρ disjoint, connected $\pi_1(S)$-sets. Take $e_\rho \in E_\rho$,then the pointed set (E_ρ, e_ρ) determines a homomorphism (because the $\pi_1(S)$ acts trivial upon $\mathfrak{g}$):

$$\varphi_\rho : \pi_1(S) \rightarrow \mathfrak{g}$$

and as $\pi_1(S)$-sets

$$E_\rho \xrightarrow{\sim} \pi_1(S) / \mathrm{Ker}\ (\varphi_\rho) \ .$$

On the other hand this quotient is a subgroup of $\mathfrak{g}$. Therefore deg (T_ρ) = card (E_ρ) divides the order of $\mathfrak{g}$.

5.2.6. We apply lemma 5.2.5 with $S= U_\alpha = X_\alpha \mid U_o$, $T= Y_\alpha \mid U_o = Y_\alpha \mid U_\alpha$ and $\mathfrak{g}$ = F from above. Let T_ρ be the irreducible components of $T= Y_\alpha \mid U_o$, then clearly $T=\coprod_\rho T_\rho$. The extension of the function fields $R(T_\rho)$ / $R(U_\alpha)$ are Galois extensions by 5.2.5 and the degrees divide order F, hence are prime to all residue characteristics by our assumption on F (5.2.2 iii). Let Y_ρ be the normalization of S_o in the function field $R(T_\rho)$ then $Y_\alpha=\coprod_\rho Y_\rho$ and we have morphisms

$$Y_\rho \xrightarrow{\psi} X_\alpha \xrightarrow{\varphi} S_o \ .$$

Let $s \in S_o$ be a maximal point of D'_o . Let A (resp. B_ρ) be the integral closure of $\underline{O}_{S_o,s}$ in the function field $R(X_\alpha)$ of X_α (resp. $R(Y_\rho)$ of Y_ρ). The maximal point s of D'_o determines a valuation v in $R(S_o)$ and v extends to a valuation w in $R(X_\alpha)$ with valuation ring A' ($\supset$ A); let w_j^*(j= 1,...,g) be the diferent extensions of w to $R(Y_\rho)$ with valuation ring $B_{\rho,j}$. The situation may be visualized by the following diagram

$$\begin{array}{ccccc} B_\rho & \subset & B_{\rho,j} & \subset & R(Y_\rho) = R(T_\rho) \\ \cup & & \cup & & \cup \\ A & \subset & A' & \subset & R(X_\alpha) = R(U_\alpha) \\ \cup & & & & \cup \\ \underline{O}_{S_o,s} & & & \subset & R(S_o) = R(U_o) \end{array}$$

Since $R(Y_\rho) / R(X_\alpha)$ is a Galois extension, all the w_j^* have the same ramification index e (resp. degree of the residue field extension f) and we have the following well-known formula ([8], vol.1, chap.V,Th.22):

$$(R(Y_\rho) : R(X_\alpha)) = efg .$$

Since by our remark above the left hand side is non-divisible by all residue characteristics of S_o, the e and f have the same property. That means that $R(Y_\rho) \supset R(X_\alpha)$ is tamely ramified over A' ; by assumption X_α is tamely ramified over $\underline{O}_{S,s_o}$. Hence by (2.2.5) $R(Y_\rho) \supset R(S_o)$ is tamely ramified over $\underline{O}_{S,s_o}$, i.e., Y_ρ is tamely ramified over S_o with respect to D_o'. This completes the proof of part a) .

5.2.7. Proof of 5.2.4 b It follows from 5.2.5 a that the functor

$$\underline{F} \rightsquigarrow H^2(\tilde{U}_o^t, \underline{F}) ,$$

with $\underline{F}$ as in 5.2.2, is left exact. Then this functor is pro-representable, therefore going to the group F , we see that there is a pro-finite π_1^t -group π_2^t , with properties as described in 5.2.3, such that we have a functorial isomorphism

$$H^2(\tilde{U}_o^t, \underline{F}) \xrightarrow{\sim} \mathrm{Hom}_{cont,\pi_1^t}(\pi_2^t, F) .$$

Corollary 5.2.8. Assumptions as in 5.2.4. There is an exact sequence

$$o \to H^2(\pi_1^t, F) \xrightarrow{\rho} H^2(U_o, \underline{F}) \to H^2(\tilde{U}_o^t, \underline{F})^{\pi_1^t} = \left\{ \mathrm{Hom}_{cont,\pi_1^t}(\pi_2^t, F) \right\}^{\pi_1^t} .$$

Proof: Apply the spectral sequence of Hochschild-Serre (SGA 4,VIII,8.4)

$$E_2^{pq} = H^p(\pi_1^t, H^q(\tilde{U}_o^t,\underline{F})) \Rightarrow H^*(U_o,\underline{F}) .$$

Since $E^{o,1} = o$ by 5.2.5 a, we have the exact sequence

$$o \to E^{2,o} \to H^2 \to E^{o,2} .$$

Corollary 5.2.9. The element

$$k'(e) \in H^2(\pi_1^t, F) ,$$

which determines the extension of 5.1.11 (or of 5.1.5):

$$o \to K \to \pi_1^D(\mathcal{S}, \xi^*) \to \pi_1^{D'_o}(S_o, \xi_o) \to 1$$

is determined by the element $\rho(k'(e)) \in H^2(U_o,\underline{K})$, where $\underline{K}$ is the sheaf (or better the projective system of sheaves) on U_o corresponding in the sense of 5.2.3 with the π_1^t-pro-finite group K and where $\pi_1^t = \pi_1^{D'_o}(S_o,\xi_o)$. Furthermore this element $\rho(k'(e))$ has image zero in $Hom(\pi_2^t, K)$. In the following we write $k(e) = \rho(k'(e))$.

§6. Comparison of two 2-cohomology classes

6.0. Introduction. In this § we prove a theorem which is a key result for the further investigation of the exact sequence 5.1.11. Since this theorem is needed in the context of formal schemes, it is necessary to develop first the notion of étale morphism for formal schemes together with some allied notions.

In § 6 $\mathcal{S}$ denotes a locally noetherian, connected formal scheme. $\mathcal{J}$ is an Ideal of definition for $\mathcal{S}$. Put for all integers $n \geq 0$

$$S_n = (\mathcal{S}, \underline{O}_{\mathcal{S}} / \mathcal{J}^{n+1}),$$

then S_n is a locally noetherian usual scheme.

6.1. Étale morphisms of formal schemes

Definition 6.1.1. a) A morphism of locally noetherian formal schemes $f: \mathfrak{X} \to \mathcal{S}$ is said to be étale in $x \in \mathfrak{X}$ if there exists a neighbourhood V of x in $\mathfrak{X}$ and a neighbourhood U of $s = f(x)$ in $\mathcal{S}$, with $f(V) \subset U$, such that

i) $f|V: V \to U$ is adic (EGA I 10.12.1),

ii) the morphisms of usual schemes $f_n: X_n = \mathfrak{X} \times_{\mathcal{S}} S_n \to S_n$ are étale in V for all n.

b) A morphism $f: \mathfrak{X} \to \mathcal{S}$ is called étale if f is étale in every point x of $\mathfrak{X}$.

6.1.2. Remarks. a) This definition is independent of the choice of the Ideal of definition. We omit the easy proof which depends on EGA IV 17.3.3(iii).

b) An étale morphism is locally of finite type (EGA I 10.13.1 a).

c) An étale covering (3.2.2) $f: \mathfrak{X} \to \mathcal{S}$ is an étale morphism as follows by applying EGA IV 18.2.3 to the morphisms $f_n: X_n \to S_n$.

d) Let $f: X \to S$ be an étale morphism of usual schemes and T a closed subset of S. Consider the completion (EGA I 10.9.1) :

$$\hat{f}: \hat{X} = X_{/f^{-1}(T)} \to \hat{S} = S_{/T} .$$

The morphism $\hat{f}$ is étale because étaleness is preserved by the base change $S_n \to S$ for every n .

Proposition 6.1.3. If $Et(\mathcal{S})$ (resp. $Et(S_o)$) denotes the category of formal S-schemes (resp. S_o-schemes) étale over $\mathcal{S}$ (resp. over S_o) then the natural functor

$$\phi : Et(\mathcal{S}) \to Et(S_o) ,$$

defined by

$$\phi(\mathfrak{X}) = (\mathfrak{X}, \underline{O}_{\mathfrak{X}} / \mathcal{J}\underline{O}_{\mathfrak{X}}) = X_o \qquad (\mathfrak{X} \in Et(\mathcal{S})),$$

is an equivalence of categories. By this equivalence fibre products of formal schemes correspond with fibre products of usual schemes.

Similar results hold if S_o is replaced by S_n $(n > o)$

Proof: EGA IV 18.1.2, EGA I 10.12.3 and 10.7.4.

6.1.4. Descent lemma. Let $\varphi: \mathcal{S}' \to \mathcal{S}$ be an étale, quasi-compact surjective morphism of locally noetherian formal schemes. Put $\mathcal{S}'' = \mathcal{S}' \times_{\mathcal{S}} \mathcal{S}'$ and $\varphi . p_1 = \varphi . p_2 = \Phi$. Then:

i) $\underline{O}_{\mathcal{S}} \to \varphi_*(\underline{O}_{\mathcal{S}'}) \rightrightarrows \Phi_*(\underline{O}_{\mathcal{S}''})$ is exact ,

ii) φ is a morphism of effective descent for the category of coherent $\underline{O}_{\mathcal{S}}$-Modules (resp. of coherent, locally free $\underline{O}_{\mathcal{S}}$-Modules),

iii) φ is a morphism of effective descent for the category of coverings of $\mathcal{S}$, i.e., of "arrows" $f: \mathfrak{X} \to \mathcal{S}$ finite over $\mathcal{S}$,

iv) if $f: \mathfrak{X} \to \mathcal{S}$ is adic then the diagram

$$Hom_{\mathcal{S}}(\mathcal{S}, \mathfrak{X}) \to Hom_{\mathcal{S}}(\mathcal{S}', \mathfrak{X}) \rightrightarrows Hom_{\mathcal{S}}(\mathcal{S}'', \mathfrak{X})$$

is exact.

Proof: First we note that the sequence

$$\mathcal{S} \xleftarrow{\varphi} \mathcal{S}' \leftleftarrows \mathcal{S}' \times_{\mathcal{S}} \mathcal{S}' = \mathcal{S}'' \Lleftarrow \mathcal{S}'''$$

is the inductive limit, in the sense of EGA I 10.12.3, of the sequence

$$S_n \xleftarrow{\varphi_n} S'_n \leftleftarrows S'_n \times_{S_n} S'_n = S''_n \Lleftarrow S'''_n \ ,$$

where $S'_n = \mathcal{S}' \times_{\mathcal{S}} S_n$ (see EGA I page 209 proof of ii)). Note also that $\mathcal{S}' \times_{\mathcal{S}} \mathcal{S}'$ is locally noetherian (EGA I 10.13.5) because φ is locally of finite type (6.1.2) .

Assertion i) follows from the corresponding statement for the φ_n after taking the projective limit, ii) follows for coherent sheaves from EGAI 10.11.3 and the corresponding statement for φ_n . In case of locally free $\underline{O}_{\mathcal{S}}$-Modules one uses [3], Alg. Comm.,chap. III, Th.1 §5. Next iii) follows from ii) applied on the $\underline{O}_{\mathcal{S}}$ -Algebra $f_*(\underline{O}_{\mathfrak{X}})$. Finally iv) follows from EGA I 10.12.3 and the corresponding statement for usual schemes applied to $\varphi_n (n \geqslant o)$.

<u>6.1.5."Etale topology" on $\mathcal{S}$</u>. It would be possible to develop now <u>the "étale topology" on $\mathcal{S}$</u>. However it follows from 6.1.3 that we can also work exclusively <u>with the étale topology on S_o</u> (cf. also with SGA 4 VIII section 1) and this we prefer to do. This has the advantage that we do work in a <u>familiar context as far as the topology is concerned</u>, but the drawback is that we <u>have to consider sheaves</u> in the étale topology of S_o (and also in the Zariski topology) <u>which are rather unusual as sheaves on S_o</u> (but natural on $\mathcal{S}$). We are going to list below some results needed later on.

<u>6.1.6. Some sheaves in the étale topology of S_o</u>. (In the following ϕ denotes the functor from 6.1.3) .

a) Let $\mathfrak{X} \in \mathrm{Et}(\mathcal{S})$. Then $\mathfrak{X}$ defines a sheaf (still denoted by $\mathfrak{X}$) in the étale topology of S_o; the value on $S'_o \in \mathrm{Et}(S_o)$ is given by

$$\mathfrak{X}[S'_o] = X_o(S'_o) = \mathrm{Hom}_{S_o}(S'_o, X_o) = \mathrm{Hom}_{\mathcal{S}}(\phi^{-1}(S_o), \mathfrak{X}) \ .$$

<u>Example</u>. $\mu_{\underline{n},\mathcal{S}}$ (see 3.1.8), the n_i prime to the residue characteristics.

b) The sheaf $\mathcal{G}_{m,\mathcal{S}}$ (both in the étale and in the Zariski topology of S_o): For $S'_o \in \mathrm{Et}(S_o)$ (resp. S'_o open on S_o), put $\mathcal{S}' = \phi^{-1}(S'_o)$ and

$$\mathcal{G}_{m,\mathcal{S}}[S_o'] = \Gamma(\mathcal{S}', \underline{O}^*_{\mathcal{S}'}) .$$

The fact that this is a sheaf follows from 6.1.4 i) and from $\phi(\mathcal{S}' x_{\mathcal{S}} \mathcal{S}') = S_o' x_{S_o} S_o'$.

c) Kummer sequence. If n is an integer prime to the residue characteristics then the following sequence is exact in the étale topology on S_o :

$$o \rightarrow \mu_{n,\mathcal{S}} \rightarrow \mathcal{G}_{m,\mathcal{S}} \xrightarrow{n} \mathcal{G}_{m,\mathcal{S}} \rightarrow o .$$

Proof: SGA 1 IX 3.2.

d) $H^1(S_{o,Zar.}, \mathcal{G}_{m,\mathcal{S}}) \xrightarrow{\sim} \mathrm{Pic}(\mathcal{S})$.

$H^1(S_{o,Et}, \mathcal{G}_{m,\mathcal{S}}) \xrightarrow{\sim} \mathrm{Pic}(\mathcal{S})$.

Proof: By the definition of $\mathcal{G}_{m,\mathcal{S}}$ in the Zariski topology on S_o, one has for $H^1(S_{o,Zar}, \mathcal{G}_{m,\mathcal{S}}) \xrightarrow{\sim} H^1(\mathcal{S}_{Zar}, \mathcal{G}_{m,\mathcal{S}})$ and this is isomorphic to $\mathrm{Pic}(\mathcal{S})$ (EGA O_I 5.4.7). For the étale topology the proof is the same as for usual schemes (see SGA 4 IX 3.3 or [2], chap. IV 1.2); the essential point in the proof is the descent lemma 6.1.4ii).

6.1.7. Finally we prove the following lemma (to be used in 6.2):

Lemma: Let $f: \mathcal{S}' \rightarrow \mathcal{S}$ be an étale morphism, then f is flat.

Proof: We can assume $\mathcal{S}' = \mathrm{Spf}\, A'$, $\mathcal{S} = \mathrm{Spf}\, A$ with A (resp. A') a J-adic (resp. $J' = J.A'$-adic) ring. Let $s' \in \mathcal{S}$ correspond with the prime ideal $\underline{p}'$ of A' and $s = f(s')$ with $\underline{p}$ of A. Consider the multiplicative system $T = A - \underline{p}$ in A (resp. $T' = A' - \underline{p}'$ in A') and intoduce the local rings $A\{T^{-1}\}$ resp. $A'\{T'^{-1}\}$ (see EGA O_I 7.6) . We have

$$A\{T^{-1}\} = \varprojlim_n (A/J^n)_{\underline{p}}$$

and similar for $A'\{T'^{-1}\}$. From the morphism $A \rightarrow A'$ we get a continuous homomorphism

$$\varphi: A\{T^{-1}\} \rightarrow A'\{T'^{-1}\}$$

defined as projective limit of

$$\varphi_n : (A/J^n)_{\underline{p}} \to (A'/J'^n)_{\underline{p}'} \quad .$$

By assumption $A/J^n \to A'/J'^n$ is étale for each n; hence φ_n is flat. The flatness of φ itself follows from the well-known criterion in [3], Alg. Comm. III, Th.1 §5 (which can be applied, due to ibid, §5, prop.2).

6.2. Coverings of Kummer type

6.2.1. The assumptions are as in 6.0. Let moreover $\mathcal{J}$ be an invertible Ideal on $\mathcal{S}$.

Definition: A couple $(\mathcal{Y}, \mathcal{M}_{n,\mathcal{S}})$ consisting of a covering $f: \mathcal{Y} \to \mathcal{S}$, on which $\mathcal{M}_{n,\mathcal{S}}$ operates over $\mathcal{S}$, is called a covering of $\mathcal{S}$ of Kummer type relative to $\mathcal{J}$ if there exists a covering family $\{\mathcal{S}_\alpha \to \mathcal{S}\}$ in the Zariski topology on $\mathcal{S}$ and sections $a_\alpha \in \Gamma(\mathcal{S}_\alpha, \underline{O}_{\mathcal{S}})$ with $(a_\alpha).\underline{O}_{\mathcal{S}} = \mathcal{J}|\mathcal{S}_\alpha$ such that

$$(\mathcal{Y}_\alpha = \mathcal{Y}|\mathcal{S}_\alpha, \mathcal{M}_{n,\mathcal{S}})$$

is a Kummer covering of $\mathcal{S}_\alpha$ relative to the section a_α (see 3.1.9).

Remarks: a) Morphisms of coverings of Kummer type are defined as $\mathcal{M}_{n,\mathcal{S}}$ - $\mathcal{S}$ -morphisms of the coverings.

b) By lemma 6.1.7 the notion is stable by étale base change (and more generally: by base change for which $\mathcal{J}$ remains invertible)

6.2.2. Consider in the étale topology on S_o the fibred category $C(\mathcal{J})$ (called the category of coverings of Kummer type relative to $\mathcal{J}$) defined as follows:

for $S'_o \in Et(S_o)$ put $\mathcal{S}' = \phi^{-1}(S'_o)$ (see 6.1.3) and let $C(\mathcal{J})(S'_o)$ be the category of coverings of $\mathcal{S}'$ of Kummer type relative to the inverse image $\mathcal{J}'$ on $\mathcal{S}'$ of $\mathcal{J}$ on $\mathcal{S}$.

By remark b) of 6.2.1 we see that $C(\mathcal{J})$ is a fibred category. By abuse of language we write sometimes $C(\mathcal{J})(\mathcal{S}')$ instead of $C(\mathcal{J})(S'_o)$

Our main objective in this section is:

Proposition 6.2.3. The fibred category $C(\mathcal{J})$ of coverings of Kummer type relative to $\mathcal{J}$ is a gerbe (see [4]) for the étale topology on S_o with lien $\mu_{n,\mathcal{S}}$. Moreover, the 2-cohomology class

$$c(\mathcal{J}) \in H^2(S_o, \mu_{n,\mathcal{S}})$$

determined by $C(\mathcal{J})$ is the image of $\mathcal{J}$, considered as element of $H^1(S_o, \mathcal{G}_{m,\mathcal{S}})$ (see 6.1.6 d), by the co-boundary operator

$$\partial : H^1(S_o\, \mathcal{G}_{m,\mathcal{S}}) \rightarrow H^2(S_o, \mu_{n,\mathcal{S}}),$$

obtained from the Kummer sequence (6.1.6 c)

$$o \rightarrow \mu_{n,\mathcal{S}} \rightarrow \mathcal{G}_{m,\mathcal{S}} \xrightarrow{n} \mathcal{G}_{m,\mathcal{S}} \rightarrow o$$

in the étale topology on S_o .(A more precise notation would be $C^{(n)}(\mathcal{J})$ and $c^{(n)}(\mathcal{J})$.)

Remark: If D is a positive divisor on $\mathcal{S}$ and $\mathcal{J}(D)$ the corresponding Ideal then we write C(D) (resp. c(D) instead of $C(\mathcal{J}(D))$ (resp.$c(\mathcal{J}(D))$).

The proof of 6.2.3 is preceded by several lemmas:

Lemma 6.2.4. Let $f: \mathcal{Y} \rightarrow \mathcal{S}$ be a covering. Equivalent conditions:

a) $(\mathcal{Y}, \mu_{n,\mathcal{S}})$ is a covering of $\mathcal{S}$ of Kummer type relative to $\mathcal{J}$.

b) The $\underline{O}_{\mathcal{S}}$-Algebra $f_*(\underline{O}_{\mathcal{Y}}) = \mathcal{A}$ has a $\mu_{n,\mathcal{S}}$-graduation:

$$\mathcal{A} = \bigoplus_{\alpha \in \mathbb{Z}_n} \mathcal{A}_\alpha ,$$

such that for the $\underline{O}_{\mathcal{S}}$-Modules $\mathcal{A}_\alpha$ we have:

i) $\mathcal{A}_\alpha$ is locally free of rank one $(1 \leqslant \alpha < n)$; put $\mathcal{A}_1 = \mathcal{L}$,

ii) the natural homomorphism $\mathcal{L}^\alpha \rightarrow \mathcal{A}_\alpha$ is an isomorphism $(1 \leqslant \alpha < n)$,

iii) the natural homomorphism $\rho : \mathcal{L}^n \rightarrow \mathcal{A}_o = \underline{O}_{\mathcal{S}}$ is an isomorphism $\rho : \mathcal{L}^n \xrightarrow{\sim} \mathcal{J} \subset \underline{O}_{\mathcal{S}}$.

Proof: This is a local question; the lemma follows from 1.2.5 and 3.1.9.

Corollary 6.2.5. Let ($\mathcal{Y}$, $\mu_{n,\mathcal{S}}$) be a covering of $\mathcal{S}$ of Kummer type relative to $\mathcal{J}$ and ($\mathcal{L}$, ρ) the couple described in 6.2.4. Then, with variable formal $\mathcal{S}$ -scheme g: $\mathcal{S}' \to \mathcal{S}$, we have

$$\mathcal{Y}(\mathcal{S}') \overset{\text{def}}{=\!=} \operatorname{Hom}_{\mathcal{S}}(\mathcal{S}', \mathcal{Y}) = \operatorname{Hom}_{\underline{O}_{\mathcal{S}}, \rho}(\mathcal{L}, g_*(\underline{O}_{\mathcal{S}'})),$$

where the subscript ρ means: those $\underline{O}_{\mathcal{S}}$ -homomorphisms which are compatible with $\rho : \mathcal{L}^n \to \underline{O}_{\mathcal{S}}$.

Proof: By EGA III 4.8.8 we have (see also 3.1.6):

$$\mathcal{Y}(\mathcal{S}') = \operatorname{Hom}_{\underline{O}_{\mathcal{S}}\text{-Algebras}}(f_*(\underline{O}_{\mathcal{Y}}), g_*(\underline{O}_{\mathcal{S}'}))$$

and by 6.2.4

$$\operatorname{Hom}_{\underline{O}_{\mathcal{S}}\text{-Algebras}}(f_*(\underline{O}_{\mathcal{Y}}), g_*(\underline{O}_{\mathcal{S}'})) = \operatorname{Hom}_{\underline{O}_{\mathcal{S}}, \rho}(\mathcal{L}, g_*(\underline{O}_{\mathcal{S}'})) .$$

Lemma 6.2.6. Let ($\mathcal{Y}_i$, $\mu_{n,\mathcal{S}}$) (i= 1,2) be two coverings of Kummer type relative to $\mathcal{J}$; let ($\mathcal{L}_i$, ρ_i) be the corresponding couples in the sense of 6.2.4. Then we have:

$$\operatorname{Hom}_{\mathcal{S}, \mu_{n,\mathcal{S}}}(\mathcal{Y}_1, \mathcal{Y}_2) \xrightarrow{\sim} \operatorname{Hom}_{\underline{O}_{\mathcal{S}}, \rho_1, \rho_2}(\mathcal{L}_2, \mathcal{L}_1) ,$$

where the subscript ρ_1, ρ_2 means: those homomorphisms compatible with ρ_1 and ρ_2. Moreover every such morphism is an isomorphism.

Proof: By EGA III 4.8.8 and 6.2.4 we have

$$\operatorname{Hom}_{\mathcal{S}, \mu_{n,\mathcal{S}}}(\mathcal{Y}_1, \mathcal{Y}_2) \xrightarrow{\sim} \operatorname{Hom}_{\underline{O}_{\mathcal{S}}, \mu_{n,\mathcal{S}}}(f_{2*}(\underline{O}_{\mathcal{Y}_2}), f_{1*}(\underline{O}_{\mathcal{Y}_1})) \xrightarrow{\sim} \operatorname{Hom}_{\underline{O}_{\mathcal{S}}, \mu_{n,\mathcal{S}}, \rho_2}(\mathcal{L}_2, f_{1*}(\underline{O}_{\mathcal{Y}_1})).$$

Now if we consider such a homomorphism Zariski locally on $\mathcal{S}$, then both $\mathcal{L}_1$ and $\mathcal{L}_2$ are free $\underline{O}_{\mathcal{S}}$-Modules with generators t_1 and t_2 say. If φ is such a homomorphism then φ is determined by the expression

$$\varphi(t_2) = \sum_{i=o}^{n-1} b_i \, t_1^i \qquad \text{(the } b_i \text{ are local sections in } \underline{O}_{\mathcal{S}} \text{)} .$$

Using the fact that φ is compatible with the action of $\mu_{n,\mathcal{S}}$ we obtain (cf. 1.3.11):

(1) $$b_i = o \qquad (i \neq 1) .$$

Using the fact that φ is compatible with ρ_2 we get

(2) $$\rho_2(t_2^n) = b_1^n \rho_1(t_1^n).$$

Since both $\rho_i(t_i^n)$ (i= 1,2) generate $\mathcal{J}$ we have b_1^n, hence b_1, a unit. From (1) we see that in fact φ corresponds with a homomorphism $\mathcal{L}_2 \to \mathcal{L}_1$ which is, by (2), compatible with ρ_1 and ρ_2 and is an isomorphism because b_1 is a unit.

Corollary 6.2.7. There is an equivalence between the category $C(\mathcal{J})(\mathcal{S})$ of coverings of $\mathcal{S}$ of Kummer type relative to $\mathcal{J}$ and the category of couples $(\mathcal{L}, \rho)$ as described in 6.2.4 with as morphisms such $\underline{O}_{\mathcal{S}}$-morphisms of the Modules which are compatible with the ρ's

Proof: Combine 6.2.4 and 6.2.6.

Corollary 6.2.8. The assumptions are the same as in 6.2.6. Consider the couple $(\mathcal{L}_1^{-1} \otimes \mathcal{L}_2, \rho)$, where $\rho : (\mathcal{L}_1^{-1} \otimes \mathcal{L}_2)^n \rightrightarrows \mathcal{J}^{-1} \otimes \mathcal{J} \rightrightarrows \underline{O}_{\mathcal{S}}$ is the isomorphism determined by ρ_1 and ρ_2, and let $(\mathcal{Y}_3, \mu_{n,\mathcal{S}})$ be the covering of Kummer type, relative to $\underline{O}_{\mathcal{S}}$, corresponding with this couple. Consider on $Et(S_o)$ the functor

$$T = \underline{\mathrm{Isom}}_{\mathcal{S}, \mu_{n,\mathcal{S}}}(\mathcal{Y}_1, \mathcal{Y}_2)$$

defined by the formula

$$T(S_o') = \underline{\mathrm{Isom}}_{\mathcal{S}', \mu_{n,\mathcal{S}'}}((\mathcal{Y}_1)_{\mathcal{S}'}, (\mathcal{Y}_2)_{\mathcal{S}'}) \quad ,$$

where $\mathcal{S}' = \phi^{-1}(S_o')$ (see 6.1.3). This functor is a sheaf on $Et(S_o)$ and is canonically isomorphic with the sheaf corresponding with $\mathcal{Y}_3$ (in the sense described in 6.1.6a). Furthermore this sheaf is a $\mu_{n,\mathcal{S}}$-torsor. In particular: $\underline{\mathrm{Aut}}_{\mathcal{S}, \mu_{n,\mathcal{S}}}(\mathcal{Y})$ is canonically isomorphic with $\mu_{n,\mathcal{S}}$.

Proof: In view of remark b of 6.2.1 we can assume that $S_o' = S_o$, i.e., $\mathcal{S}' = \mathcal{S}$. By 6.2.6 we have

$$\mathrm{Isom}_{\mathcal{S},\, \mu_{n,\mathcal{S}}}(\mathcal{Y}_1, \mathcal{Y}_2) \simeq \mathrm{Hom}_{\underline{O}_{\mathcal{S}},\, \rho_1,\, \rho_2}(\mathcal{L}_2, \mathcal{L}_1) .$$

Consider the natural map

$$(*) \qquad \mathrm{Hom}_{\underline{O}_{\mathcal{S}},\, \rho_1,\, \rho_2}(\mathcal{L}_2, \mathcal{L}_1) \xrightarrow{\lambda} \mathrm{Hom}_{\underline{O}_{\mathcal{S}},\, \rho}(\mathcal{L}_1^{-1} \otimes \mathcal{L}_2, \underline{O}_{\mathcal{S}})$$

defined as follows: let

$$\varphi : \mathcal{L}_2 \to \mathcal{L}_1 ,$$

then $\lambda(\varphi)$ is given by the formula

$$\check{\ell}_1 \otimes \ell_2 \mapsto \check{\ell}_1(\varphi(\ell_2)) .$$

We claim: <u>(*) is an isomorphism</u>. This is eaily checked in a straight forward manner. Next we remark that by 6.2.5 the right hand side of (*) is precisely $\mathcal{Y}_3(\mathcal{S})$. The map λ is natural, therefore it behaves functorial by morphisms $\mathcal{S}'' \to \mathcal{S}'$ and this completes the proof for the first part of 6.2.8. The assertion that $\mathcal{Y}_3$ is a $\mu_{n,\mathcal{S}}$-torsor can be proved locally, i.e., we can assume $\mathcal{L}_i \simeq \underline{O}_{\mathcal{S}}$ (i= 1,2) . Let $\rho(1) = u \in \Gamma(\mathcal{S}, \underline{O}_{\mathcal{S}})$, then u is a unit. $\mathcal{Y}_3$ is then a Kummer covering of $\mathcal{S}$, relative to the unit u, and the assertion follows (going from $\mathcal{S}$ = Spf A to Spec A) from 1.4.5 (or is obtained by a direct check.) In case $\mathcal{Y}_1 = \mathcal{Y}_3 = \mathcal{Y}$ we have u= 1 and $\mathcal{Y}_3 = \mu_{n,\mathcal{S}}$.

<u>Corollary 6.2.9</u>. The assumptions are as in 6.2.6. Then there exists for every $s \in S_o$ an étale neighbourhood S_o' of s in S_o such that for $\mathcal{S}' \in \mathrm{Et}(\mathcal{S})$, with $\phi(\mathcal{S}') = S_o'$ (see 6.1.3), we have

$$(\mathcal{Y}_1)_{\mathcal{S}'} \simeq (\mathcal{Y}_2)_{\mathcal{S}'} .$$

<u>Proof:</u> This follows from 6.2.8. We can find an étale neighbourhood S_o' for $s \in S_o$ such that for $\mathcal{S}' = \phi^{-1}(S_o')$ we have $\mu_{n,\mathcal{S}}(\mathcal{S}') \neq \emptyset$.

<u>6.2.10. Proof of 6.2.3</u>. First we want to see that in $C(\mathcal{Y})$ the descent theorems hold. By 6.1.4 iii) we have descent in the category

of coverings of $\mathcal{S}$ with respect to surjective étale morphisms $\mathcal{S}' \to \mathcal{S}$ (i.e., with respect to étale morphisms $S_0' \to S_0$). It follows also by 6.1.4 iii) that $\mu_{n,\mathcal{S}'}$-morphisms give after descent $\mu_{n,\mathcal{S}}$-morphisms. There remains to be seen that a covering of Kummer type of $\mathcal{S}'$ relative to $\mathcal{J}'$ gives after descent a covering of Kummer type of $\mathcal{S}$ relative to $\mathcal{J}$. For this we can assume that $\mathcal{S}$ = Spf A; consider the corresponding covering of Spec A= S, i.e., we are reduced to the case of usual coverings. We use the characterization 1.2.5 for Kummer coverings; the assertion follows because the condition b) of 1.2.5 is stable by faithfully flat descent.

The fact that $C(\mathcal{J})$ is a gerbe follows from 6.2.9, 6.2.6 and from the observation that we can find a Zariski covering $\{\mathcal{S}_\alpha \to \mathcal{S}\}$ such that $C(\mathcal{J})(\mathcal{S}_\alpha) \neq \emptyset$. By 6.2.8 the lien is $\mu_{n,\mathcal{S}}$.

Finally the assertion: In $H^2(S_0, \mu_{n,\mathcal{S}})$ we have $c(\mathcal{J}) = \partial(\mathcal{J})$. According to "general facts from 2-cohomology" (see [4]) $\partial(\mathcal{J})$ is the cohomology class determined by the gerbe consisting of the couples $(\mathcal{L}', \rho')$ on $\mathcal{S}'$ (étale over $\mathcal{S}$) with $\mathcal{L}'$ an invertible $\mathcal{O}_{\mathcal{S}'}$-Module and $\rho': \mathcal{L}'^n \xrightarrow{\sim} \mathcal{J}'$ (where $\mathcal{J}'$ is the inverse image of $\mathcal{J}$ by $\mathcal{S}' \to \mathcal{S}$). However by 6.2.7 this gerbe is precisely the gerbe $C(\mathcal{J})$.

6.3. Comparison of two 2-cohomology classes

6.3.1. The assumptions and notations are as in 6.0. Let furthermore H be an ordinary finite group and $\Phi: \mathcal{P} \to \mathcal{S}$ a H-torsor in RevEt($\mathcal{S}$). We recall that this means that Φ is a surjective étale covering and the morphism $\mathcal{P} \times_{\mathcal{S}} H_{\mathcal{S}} \to \mathcal{P} \times \mathcal{P}$, given by $(p,h) = (p, ph)$, with $H_{\mathcal{S}}$ the constant group H, is an isomorphism. By the conventions of 6.1.6 a) we consider $\mathcal{P}$ as a sheaf in the étale topology on S_0 and as such it is again a H-torsor, where now H is considered as the constant group of $Et(S_0)$.

Consider on the one hand the category of finite H-sets, with H operating on the left, and on the other hand the category of sheaves for the étale topology on S_o. Then there is a "morphism in the sense of topi" :

(topos of sheaves on $Et(S_o)$) $\to$ (topos of finite H-sets),

such that the <u>inverse image functor</u> Ψ can be described as follows: Let E be a left H-set and $E_{\mathcal{J}}$ the corresponding formal $\mathcal{J}$-scheme

$$E_{\mathcal{J}} = \coprod_{E} \mathcal{J},$$

then

$$E \overset{\Psi}{\rightsquigarrow} E^{\mathcal{P}} = \mathcal{P} \times^{H} E_{\mathcal{J}},$$

where the $E^{\mathcal{P}} \in \mathrm{RevEt}(\mathcal{J})$ is considered (6.1.6a) as a sheaf on $Et(S_o)$. We recall that $E^{\mathcal{P}}$ is obtained from $E_{\mathcal{J}}$ by "twisting" with $\mathcal{P}$ (cf. Serre, Cohomologie Galoisienne, p I-59), namely as quotient of $\mathcal{P} \times_{\mathcal{J}} E_{\mathcal{J}}$ under the action of H by means of the formula $(p,e).h = (ph, h^{-1}e)$. We also remark that if we take a pointing in $\mathcal{P}(\xi_o)$, with ξ_o a geometric point on $\mathcal{J}$, then we get a homomorphism $\pi_1(\mathcal{J}, \xi_o) \to H$. Hence E becomes a $\pi_1(\mathcal{J}, \xi_o)$-set and therefore determines an étale covering $\mathcal{E}$ of $\mathcal{J}$ (cf. also 5.2.2); now $E^{\mathcal{P}}$ and $\mathcal{E}$ are $\mathcal{J}$-isomorphic. This follows immediately by looking to the operation of $\pi_1(\mathcal{J}, \xi_o)$ on a geometric fibre.

Finally we remark that, from the functor Ψ above, we get homomorphisms of the cohomology groups (in case E is a H-group):

$$H^i(H,E) \to H^i(S_o, E^{\mathcal{P}}).$$

<u>6.3.2.</u> Let

(e) $$o \to F \to G \to H \to 1$$

be a sequence of ordinary finite groups with F abelian, hence H operates on F via inner automorphisms of G. The above extension is determined by an element

$$k(e) \in H^2(H,F).$$

By the above homomorphism of 6.3.1 we get an element, still denoted by the same letter:

$$k(e) \in H^2(S_0, F^{\mathcal{P}}) .$$

Moreover by the general 2-cohomology theory of [4], k(e) is the class of a gerbe K(e) consisting of the couples $(\mathcal{P}^+, \lambda)$ with $\mathcal{P}^+$ a G-torsor for the étale topology on S_0 and $\lambda : \mathcal{P}^+ \to \mathcal{P}$ a morphism compatible with the action of G on $\mathcal{P}^+$ and on $\mathcal{P}$ (via the action of H).

6.3.3. On top of the assumptions in 6.3.1 and 6.3.2 we assume that we have a morphism φ of formal $\mathcal{S}$-schemes as follows:

$$\mathcal{R} \xrightarrow{\varphi} \mathcal{P} \xrightarrow{\psi} \mathcal{S}$$

such that:

a) φ is finite (ψ is also finite by the assumption in 6.3.1),

b) G operates on $\mathcal{R}$; by 6.3.1 H operates on $\mathcal{P}$ and $\mathcal{P}$ is a H-torsor ; both operations are on the right, and are operations over $\mathcal{S}$,

c) φ is a G-morphism (G operates via H on $\mathcal{P}$) ,

d) there exists, for a certain n prime to the residue characteristics, an isomorphism in RevEt($\mathcal{S}$)

$$i: \mu_{n,\mathcal{S}} \xrightarrow{\sim} F^{\mathcal{P}} ,$$

e) there exists an invertible Ideal $\mathcal{J}$ on $\mathcal{S}$ such that the inverse image $\psi^{-1}(\mathcal{J}) = \mathcal{J}'$ on $\mathcal{P}$ is again invertible and such that $(\mathcal{R}, \mu_{n,\mathcal{P}})$ is a covering of $\mathcal{P}$ of Kummer type (6.2.1) relative to $\mathcal{J}'$.

Note that by the isomorphism i lifted to $\mathcal{P}$:

$$i_{\mathcal{P}} : \mu_{n,\mathcal{P}} \xrightarrow{\sim} F ,$$

we have indeed that $\mu_{n,\mathcal{P}}$ operates on $\mathcal{R}$ over $\mathcal{P}$, therefore assumption e) makes sense.

6.3.4. Using the invertible Ideal $\mathcal{J}$ on $\mathcal{S}$ we have an element (6.2.3)

$$c(\mathcal{J}) \in H^2(S_0, \mu_{n,\mathcal{S}}) .$$

From the extension (e) in 6.3.2 we have

$$k(e) \in H^2(S_0, F^{\mathcal{P}}).$$

Furthermore from the isomorphism i in 6.3.3 d we have an isomorphism

$$H^2(S_o,i) : H^2(S_o, \mu_{n,\mathcal{S}}) \xrightarrow{\sim} H^2(S_o,F^{\mathcal{P}}) .$$

<u>Theorem 6.3.5</u>. With the above assumptions and notations, and identifying the cohomology groups $H^2(S_o, \mu_{n,\mathcal{S}})$ and $H^2(S_o,F^{\mathcal{P}})$ by means of $H^2(S_o,i)$, we have

$$k(e) = c(\mathcal{J}) .$$

<u>Proof:</u> Consider the gerbe $C = C(\mathcal{J})$ (6.2.3) and the gerbe $K = K(e)$ (6.3.2). We want to find a functor

$$\lambda : C \to K$$

such that for the corresponding morphism of liens

$$j: \mu_{n,\mathcal{S}} \to F^{\mathcal{P}}$$

we have

$$j = i \quad .$$

Let $(S_o)_* \to S_o$ be étale; we are going to define $\lambda(S_{o_*})$. Put $\mathcal{S}_* = \phi^{-1}(S_{o_*})$ (see 6.1.3), i.e., $\mathcal{S}_* \to \mathcal{S}$ is étale. Denote by $\mathcal{R}_*$, $\mathcal{P}_*$ (resp. $\mathcal{J}_*$) the formal schemes (resp. Ideal) obtained by base change $\mathcal{S}_* \to \mathcal{S}$.

Start with $\mathcal{Z} \in C(\mathcal{J})(S_{o_*})$, i.e., $\mathcal{Z}$ is a covering of $\mathcal{S}_*$ of Kummer type relative to $\mathcal{J}_*$. We have the following situation:

$$\begin{array}{ccccc}
\mathcal{Z} & \longleftarrow & \mathcal{Z}_{\mathcal{P}} & & \\
\downarrow & & \downarrow & & \\
\mathcal{S}_* & \longleftarrow & \mathcal{P}_* & \longleftarrow & \mathcal{R}_* \\
\downarrow & & \downarrow & & \downarrow \\
\mathcal{S} & \xleftarrow{\ \psi\ } & \mathcal{P} & \xleftarrow{\ \varphi\ } & \mathcal{R}
\end{array}$$

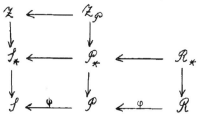

Now $\mu_{n,\mathcal{S}}$ operates on $\mathcal{Z}$ (write for this time the operation on the left), hence on $\mathcal{Z}_{\mathcal{P}}$. Furthermore the twisted sheaf $F^{\mathcal{P}}$ operates on the left on $\mathcal{R}$ in a way compatible with the operation of G on the right (set theoretically this operation is given by the formula $\overline{(p,f)}.r = r.f^{-1}$, where the representative (p,f) is chosen in such a way that $\varphi(r)= p$). Using the isomorphism i from 6.3.3 d we have that $\mathcal{M}_{n,\mathcal{S}}$ operates on the left on $\mathcal{R}$ in a way compatible with the operation by G on the right.

Consider now the functor

$$\lambda(S_{o_*}): C(\mathcal{J})\ (S_{o_*}) \to K(e)\ (S_{o_*})$$

defined as follows:

$$\lambda: \mathcal{Z} \rightsquigarrow (\mathcal{Q} = \underline{\mathrm{Isom}}_{\mathcal{P}_*, \mathcal{M}_{n,\mathcal{S}_*}}(\mathcal{R}_*, \mathcal{Z}_{\mathcal{P}}) ,q)$$

By 6.2.8 the right hand side may be considered as an étale covering of $\mathcal{P}_*$, therefore we have a natural morphism of formal $\mathcal{S}_*$-schemes

$$q: \mathcal{Q} \to \mathcal{P}_* ,$$

and both $\mathcal{Q}$ and $\mathcal{P}_*$ may be considered as sheaves in the étale topology on S_{o_*} (cf.6.1.6a). There remain to be shown the following points:

I) $(\mathcal{Q}, q: \mathcal{Q} \to \mathcal{P}_*)$ is in $K(e)(S_{o*})$.

Proof: G operates over $\mathcal{S}_*$ on the right of $\mathcal{R}_*$ and on $\mathcal{Z}_{\mathcal{P}}$ (via the operation on $\mathcal{P}$) and this operation is compatible with the operation on the left by $\mathcal{M}_{n,\mathcal{S}_*}$. It follows that G operates on $\mathcal{Q}$ "in the usual way" by the formula

(*) $\qquad (\alpha.g)\ (r)= [\alpha(r.g^{-1})].g \qquad (g\in G,\ \alpha\in \mathcal{Q}\ ,\ r\in \mathcal{R}_*)$;

this operation is compatible with the operation of $\mathcal{M}_{n,\mathcal{S}_*}$ and q is a G-morphism. In order to see that $\mathcal{Q}$ is a G-torsor over $\mathcal{S}_*$ it suffices to see that $\mathcal{Q}$ is a F-torsor over $\mathcal{P}_*$. The operation of F on $\mathcal{Q}$ over $\mathcal{P}_*$ is given by the above formula, i.e., by

(* *) $\qquad (\alpha.f)\ (r)= \alpha(r.f^{-1})$,

because the operation of F on $\mathcal{Z}_{\mathcal{P}}$ is trivial. The fact that

$$\mathcal{Q} \times F \to \mathcal{Q} \times \mathcal{Q}$$

is a bijective map (after evaluation on $Et(P_o)$) is seen by straight forward calculation using (* *). From this and from the fact that -as formal schemes- $\mathcal{Q}$ is an étale covering of $\mathcal{P}_*$ we get that $\mathcal{Q}$ is a F-torsor over $\mathcal{P}_*$.

II) The functor $\lambda(S_{o_*})$ behaves correctly by base change in $Et(S_o)$.

We omit the proof of this.

III) From the functor λ we get a homomorphism of groups

$$j: \mathcal{M}_{n,\mathcal{S}} \to F^{\mathcal{P}} = \underline{\mathrm{Aut}}\,(\mathcal{Q}, q).$$

We want to show that

$$j = i$$

<u>Proof:</u> Since it suffices to prove this locally we can assume that we have sections $\varepsilon \in \mathcal{M}_{n,\mathcal{S}}(\mathcal{S}_*)$ and $\alpha \in \mathcal{Q}\,(\mathcal{S}_*)$. The section α gives via $q: \mathcal{Q} \to \mathcal{P}_*$ an element $p \in \mathcal{P}(\mathcal{S}_*)$ and using this we can (over $\mathcal{S}_*$) identify $F^{\mathcal{P}}$ and F.

By the definition of the functor λ we have

$$j(\varepsilon)\,(\alpha) = \varepsilon.\alpha$$

Since α is a $\mathcal{M}_{n,\mathcal{S}}$-morphism we have (point-set theoretically and with $r \in \mathcal{R}_*$)

$$\varepsilon.\alpha(r) = \alpha(\varepsilon.r) = \alpha(i(\varepsilon).r) = \alpha(r.i(\varepsilon)^{-1}),$$

because this is the way the action of $\mathcal{M}_{n,\mathcal{S}}$ on $\mathcal{R}$ is defined. On the other hand by the formula (**) we have

$$(j(\varepsilon)\alpha)\,(r) = \alpha(r.j(\varepsilon)^{-1}).$$

Since $\mathcal{Q}$ is a F-torsor over $\mathcal{P}_*$, the element $f \in F$ which makes the diagram

$$\begin{array}{ccc} \mathcal{R}_* & \xrightarrow{\ \alpha\ } & \mathcal{Z}_{\mathcal{P}} \\ \downarrow f & & \downarrow f \quad \text{(trivial action!)} \\ \mathcal{R}_* & \xrightarrow{\ \varepsilon.\alpha\ } & \mathcal{Z}_{\mathcal{P}} \end{array}$$

commutative, is unique. Hence $i(\varepsilon) = j(\varepsilon)$, i.e., $i = j$.

§7. The tame fundamental group of a formal neighbourhood of an irreducible divisor (continued)

7.1. Determination of the extension (see 5.1.11)

7.1.1. The assumptions are the same as in 5.1.1; see also 5.2.1 for the notations. According to 5.2.9 the extension from 5.1.11:

$$o \to K \to \pi_1^D(\mathcal{S},\xi^*) \to \pi_1^{D'_o}(S_o,\xi_o) \to 1$$

is determined by an element

$$k(e) \in H^2(U_o,\underline{K}) \ .$$

Here $\underline{K}$ is a projective system of sheaves in the étale topology of U_o and is determined, in the way described in 5.2.3, by the profinite π_1^t- group K. Recall that we use the abbreviation

$$\pi_1^t = \pi_1^{D'_o}(S_o,\xi_o) \ .$$

Furthermore we have, from 5.1.11, a canonnical, continuous, surjective homomorhism

$$j: \mu^t(\xi_o) \to K \to 1 \ .$$

Hence we have a homomorphism of the corresponding projective system of sheaves

$$j: \mu^t \to \underline{K} \ ,$$

where μ^t is the projective system (filtered by the relation $n \mid n'$)

$$\mu^t = (\mu_{n,S_o}) \quad , \qquad n \text{ invertible on } S_o \ .$$

7.1.2. Consider, in the étale topology on S_o, the Kummer sequence (see 6.1.6c):

$$o \to \mu_{n,\mathcal{S}} \to \mathcal{G}_{m,\mathcal{S}} \xrightarrow{n} \mathcal{G}_{m,\mathcal{S}} \to o \ .$$

This gives an exact sequence of cohomology groups:

$$\cdots \to H^1(S_o,\mathcal{G}_m,\mathcal{S}) \xrightarrow{n} H^1(S_o,\mathcal{G}_m,\mathcal{S}) \xrightarrow{\partial} H^2(S_o,\mu_{n,\mathcal{S}}) \to \cdots$$

$$\downarrow \wr$$

$$H^2(S_o,\mu_{n,S_o}) \ .$$

From the divisor D_o on $\mathcal{S}$ we get (see 6.1.6d)

$$\text{class}\ (D_o) \in H^1(S_o,\mathcal{G}_m,\mathcal{S}) \ .$$

Let the image under ∂ be denoted by

$$c^{(n)}(D_o) \in H^2(S_o,\mu_{n,S_o}) \ .$$

Next, going to the projective system μ^t we get, with the conventions of 5.2.3 (this time on S_o itself instead of on U_o), a class

$$c(D_o) = \varprojlim_n c^{(n)}(D_o) \in H^2(S_o,\mu^t) \ .$$

Finally we write

$$c'(D_o) \in H^2(U_o,\mu^t)$$

for the image of $c(D_o)$ by the canonnical homomorphism

$$H^2(S_o,\mu^t) \to H^2(U_o,\mu^t) \ .$$

obtained from the inclusion $U_o \subset S_o$.

Theorem 7.1.3. With the above assumptions and notations, and writing

$$j_* = H^2(U_o,j) : H^2(U_o,\mu^t) \to H^2(U_o,\underline{K})$$

for the homomorphism of cohomology groups corresponding with the homomorphism (from 5.1.11)

$$j : \mu^t \to \underline{K} \ ,$$

we have the relation

$$k(e) = j_*(c'(D_o)) \ .$$

Proof: Consider the exact sequence of 5.1.11

$$(*) \qquad o \to K \to \pi_1^D(\mathcal{S},\xi^*) \to \pi_1^{D'_o}(S_o,\xi_o) \to 1 \ .$$

we have $\pi_1^D(\mathcal{S},\xi^*) = \varprojlim_\gamma \Gamma_\gamma$, with Γ_γ finite groups and similarly for

$\pi_1^{D'_0}(S_0,\xi_0)$. Let K_γ be the image of K in Γ_γ and put $\Gamma'_\gamma = \Gamma_\gamma / K_\gamma$. The above exact sequence is the projective limit of the corresponding exact sequences of finite groups:

$$(**) \qquad o \to K_\gamma \to \Gamma_\gamma \to \Gamma'_\gamma \to 1 .$$

Here Γ_γ (resp. Γ'_γ) is the Galois group of a connected, pointed object $\mathcal{X} \in \mathrm{Rev}^D(\mathcal{S})$ (resp. $\mathcal{X}' \in \mathrm{Rev}^{D'}(\mathcal{S})$) and we have morphisms

$$\mathcal{S} \xleftarrow{\psi} \mathcal{X}' \xleftarrow{\varphi} \mathcal{X} ,$$

with φ a Γ_γ- morphism.

Since Γ'_γ, and hence $\pi_1^t = \pi_1^{D'_0}(S_0, \xi_0)$, operates on K_γ we have by 5.2.2 a sheaf $\underline{K}_\gamma$ for the étale topology on U_0. On the other hand write $U = \mathcal{S}|U_0$, then clearly $\mathcal{X}'|U$ is a Γ'_γ- torsor. Therefore we have by 6.3.1 a "twisted sheaf" $(K_\gamma)^{\mathcal{X}'|U}$ for the étale topology on U_0. Since $\mathcal{X}'$ is a <u>pointed</u> object in $\mathrm{Rev}^{D'}(\mathcal{S})$, these two sheaves are canonically isomorphic

$$(K_\gamma)^{\mathcal{X}'|U} \simeq \underline{K}_\gamma$$

(see the lines just above the final remark in 6.3.1).

Next, by 5.1.7, there is for a suitable integer n, prime to the residue characteristics of $\mathcal{S}$, an isomorphism

$$j_n(\xi_0): \mu_n(\xi_0) \xrightarrow{\sim} K_\gamma .$$

Therefore both sheaves are (on U_0) isomorphic with $\mu_{n,\mathcal{S}}$; it follows also that $\mu_{n,\mathcal{S}}$ lifted to $\mathcal{X}'|U$ becomes constant and isomorphic with K_γ . Since K_γ operates on $\mathcal{X}|U$ over $\mathcal{X}'|U$ we have that $\mu_{n,\mathcal{S}'}$ operates on $\mathcal{X}|U$ over $\mathcal{X}'|U$. Now <u>we claim</u> that with respect to this operation $\mathcal{X}|U$ is a <u>covering of $\mathcal{X}'|U$ of Kummer type, relative to the Ideal $\Phi^{1}(\mathcal{J}) = \mathcal{J}'$.</u>

In order to see this it suffices to check the conditions of 6.2.4. This can be done locally on $\mathcal{X}'|U$, therefore we can reduce to usual schemes by 4.1.3 and next we can make an étale base change.

Using 2.2.5 we see that $\mathfrak{X}|U$ is tamely ramified over $\mathfrak{X}'|U$ with respect to the divisor defined by $\mathcal{Y}'$ and we can apply 2.3.4.ii). Since the degree of $\mathfrak{X}|U$ over $\mathfrak{X}'|U$ is equal to the order of the group K_γ, and since K_γ is the inertia group of a geometric point η_o in $\mathfrak{X}(\xi_o)$, we have from 2.3.4.ii) that as a scheme $\mathfrak{X}|U$ is a Kummer covering over $\mathfrak{X}'|U$ (and not a union of Kummer coverings). From the way the above isomorphism $j_n(\xi_o)$ is obtained (see the usual valuation theory) we have that the above obtained action of μ_n agrees with the action which exists implicity by the assertion 2.3.4 ii).

From the above discussion we see that we can apply 6.3.5 to the coverings

$$\mathcal{Y}|U \xleftarrow{\Phi} \mathfrak{X}'|U \xleftarrow{\varphi} \mathfrak{X}|U$$

and to the exact sequence (**). Let $k(e_\gamma)$ be the element in $H^2(U_o,\underline{K}_\gamma)$ determined by this exact sequence. Then we have by the isomorphism

$$(j_n)_* : H^2(U_o,\mu_n) \xrightarrow{\sim} H^2(U_o,\underline{K}_\gamma)$$

the relation (6.3.5):

$$k(e_\gamma) = \left\{(j_n)_*(c'^{(n)}(D_o))\right\}$$

(recall: the prime means restriction to U_o). The theorem itself follows by taking the projective limit (over γ, resp. n).

7.2. Determination of the kernel K

7.2.1. After the result 7.1.3 we see that the extension of 5.1.11 is determined as soon as we "know" the group K. For this we use the canonical, surjective, continuous homomorphism (5.1.11):

$$j : \mu^t(\xi_o) \to K ,$$

i.e., we want to determine ker(j).

With the usual interplay between profinite π_1^t- groups and projective systems of sheaves for the étale topology on U_o we have the following commutative diagram (see 5.2.8; by abuse of language we also use the letter j for the induced maps of the cohomology groups):

$$
\begin{array}{ccccccc}
o \to & H^2(\pi_1^t, \mathcal{M}^t(\xi_o)) & \to & H^2(U_o, \mathcal{M}^t) & \to & \{H^2(\vec{U}_o^t, \mathcal{M}^t)\}^{\pi_1^t} \\
& j\downarrow & & j\downarrow & & j\downarrow \\
o \to & H^2(\pi_1^t, K) & \to & H^2(U_o, \underline{K}) & \to & \{H^2(\vec{U}_o^t, \underline{K})\}^{\pi_1^t}
\end{array}
$$

and furthermore

$$
\begin{array}{ccc}
H^2(\vec{U}_o^t, \mathcal{M}^t) = & \mathrm{Hom}_{\mathrm{cont}, \pi_1^t}(\pi_2^t, \mathcal{M}^t(\xi_o)) \\
j\downarrow & j\downarrow \\
H^2(\vec{U}_o^t, \underline{K}) = & \mathrm{Hom}_{\mathrm{cont}, \pi_1^t}(\pi_2^t, K)
\end{array}
$$

The element $c'(D_o) \in H^2(U_o, \mathcal{M}^t)$ of 7.1.2 determines a continuous π_1^t- homomorphism:

$$\beta : \pi_2^t \to \mathcal{M}^t(\xi_o) .$$

By 7.1.3 and the last remark in 5.2.9 we have

$$j . \beta = o$$

i.e.,

$$(*) \qquad \ker(j) \supset \mathrm{Im}(\beta) .$$

Proposition 7.2.2. Under the usual assumptions and notations (see 5.1.1 and 5.2.1), together with the additional assumption

$$D_{i,o} \cap D_{j,o} = \emptyset \qquad (i \neq o,\ j \neq o,\ i \neq j) ,$$

we have

$$\mathrm{Im}(\beta) = \ker(j) .$$

Proof: Taking into account the inclusion $(*)$, there remains to be shown the following: if N is an open subgroup of $\mathcal{M}^t(\xi_o)$ such that

$$N \supset \mathrm{Im}(\beta) ,$$

then

$$N \supset \ker(j) .$$

By 5.1.7 it suffices to show that there exists a Galois object

$$\mathcal{Y} \in \mathrm{Rev}^{D}(\mathcal{S})$$

such that $\mu^{t}(\xi_o)/N = \mu_{n_o}(\xi_o)$ is the inertia group of a point of the geometric fibre $\mathcal{Y}_{\xi_o}$. In the following the integer n_o is kept fixed (note that n_o is prime to the residue characteristics of $\mathcal{S}$).

7.2.3. Before proceeding we claim that we can make the following additional assumptions:

a) the image of $c'(D_o)$ in $H^2(U_o, \mu_{n_o})$ is zero,

b) μ_{n_o} is constant on U_o.

In order to see this we recall that $\vec{U}_o^{t}$ is obtained (see 5.2.1) as an inductive system of $X_\alpha \in \mathrm{Rev}^{D_o'}(S_o)$. Due to the assumption $N \supset \mathrm{Im}(\beta)$, condition a) is fulfilled if we are able to replace U_o by $U_\alpha = X_\alpha | U_o$ for large α, and clearly the same is true for condition b).

With the $X_\alpha \in \mathrm{Rev}^{D_o'}(S_o)$ corresponds (4.3.2) $\mathcal{S}_\alpha \in \mathrm{Rev}^{D'}(\mathcal{S})$; let $f_\alpha : \mathcal{S}_\alpha \to \mathcal{S}$ be the structure map. Put $I' = \{i;\ i \neq o\}$, $I'' = \{o\}$; we have by assumption $D_i \cap D_j = \emptyset$ for $i \neq j$ $(i,j \in I')$. Therefore we can apply 4.4.4 to $f_\alpha : \mathcal{S}_\alpha \to \mathcal{S}$; we see that $\mathcal{S}_\alpha$ and the reduced inverse image D_* on $\mathcal{S}_\alpha$ of the divisor D on $\mathcal{S}$ fulfil the conditions of 5.1.1 (the regularity of $\mathcal{S}_\alpha$ outside the divisors follows from 3.2.3 and 3.1.3). We have by 4.4.4

$$D_* = D_*' + f_\alpha^{-1}(D_o)$$

with D_*' the inverse image of $D' = \sum_{i \neq o} D_i$.

For abbreviation put $\pi = \pi_1^{D}(\mathcal{S}, \xi_*)$, $\pi^* = \pi_1^{D_*}(\mathcal{S}_\alpha, \xi_\alpha^*)$, $\pi' = \pi_1^{D'}(\mathcal{S}, \xi)$, $(\pi^*)' = \pi_1^{D_*'}(\mathcal{S}_\alpha, \xi_\alpha)$. Then we summarize the situation in the following diagram, in which rows and colums are exact, in which the rectangle is commutative up to an inner automorphism and in which Γ_α denotes the Galois group of $\mathcal{S}_\alpha$ (or of X_α):

$$
\begin{array}{ccccccccc}
 & & & & 1 & & 1 & & \\
 & & & & \downarrow & & \downarrow & & \\
1 & \longrightarrow & K^* & \longrightarrow & \pi^* & \longrightarrow & (\pi^*)' & \longrightarrow & 1 \\
 & & & & \downarrow & & \downarrow & & \\
1 & \longrightarrow & K & \longrightarrow & \pi & \longrightarrow & \pi' & \longrightarrow & 1 \\
 & & & & \downarrow & & \downarrow & & \\
 & & & & \Gamma_\alpha & = & \Gamma_\alpha & & \\
 & & & & \downarrow & & \downarrow & & \\
 & & & & 1 & & 1 & &
\end{array}
$$

For the exactness of the columms we have used 4.4.6. From 4.4.6 follows also that π^* (resp. $\pi^{*'}$) is an open subgroup of π (resp. of π'). Therefore we have that $K \subset \pi^*$ and $K = K^*$.

Replace $\mathcal{S}$ and D by $\mathcal{S}_\alpha$ and D_*. As we have remarked above this does not destroy the assumptions made in 7.2.2. Now it suffices to show that there exists an object $\mathcal{Y} \in \mathrm{Rev}^{D_*}(\mathcal{S}_\alpha)$, which is Galois and which has inertia group $\mu_{n_0}(\xi_0)$. Because then we have a continuous <u>surjective</u> homomorphism

$$K \to \mu_{n_0}(\xi_0) \quad ,$$

and this is sufficient for the assertion that $N \supset \mathrm{Ker}(j)$.

<u>Conclusion</u>: We can make the additional assumptions a) and b) of 7.2.3.

<u>Lemma 7.2.4</u>: The assumptions are the same as in 7.2.2. Assume moreover that:

a) the image $c'^{(n)}(D_0)$ of $c'(D_0)$ in $H^2(U_0, \mu_{n_0})$ is zero,

b) μ_{n_0} is constant on U_0.

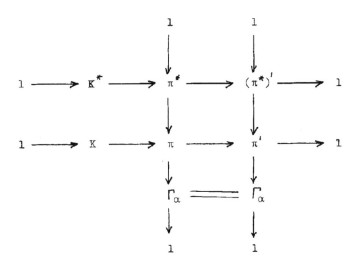

Then there exists a $f: \mathcal{T} \to \mathcal{S}$ such that $(\mathcal{T}, \mu_{n_o})$ is a covering of $\mathcal{S}$ of Kummer type relative to the divisor

$$\bar{D} = D_o + \Sigma_{i \in I'} q_i D_i \qquad (q_i \text{ positive integers}) .$$

<u>7.2.5. Proof that 7.2.4 ⇒ 7.2.2</u>. According to the remarks in 7.2.2 it suffices to construct a $\mathcal{Y} \in \mathrm{Rev}^D(\mathcal{S})$, which is Galois and such that the inertia group of a point of the fibre $\mathcal{Y}_{\xi_o}$ is $\mu_{n_o}(\xi_o)$. We proceed in several steps:

<u>Step I</u>. Assume $\mathcal{S}$ =Spf A with A a J-adic noetherian ring. Let S= Spec A, then S is regular because $\mathcal{S}$ is regular. Let $\mathcal{T}$ be as in 7.2.4, then $\mathcal{T}$ = Spf B with B a finite A-algebra. Put T= Spec B, then (T, μ_{n_o}) is a covering of Kummer type of S relative to the corresponding divisor D on S (use -for instance- 6.2.4). Let Y be the normalization of T over S and $\mathcal{Y} = Y_{/V(J)}$. <u>It suffices to prove that $Y \in \mathrm{Rev}^D(S)$ and that Y is a Galois covering with inertia group μ_{n_o} for a point $\eta_o \in Y_{\xi_o}$</u>. Because then $\mathcal{Y} \in \mathrm{Rev}^D(\mathcal{S})$, is Galois with the same group, and the inertia groups are the same for $\eta_o \in Y_{\xi_o}$ and for $\eta_o \in \mathcal{Y}_{\xi_o}$ (see 4.1.3) .

We can assume that S is connected, hence irreducible. In order to see that $Y \in \mathrm{Rev}^D(S)$, first look to T. Locally on S we have that $D_i = \mathrm{div}(t_i)$ and T= Spec B is given by $B = A[\tau]$ with

$$(*) \qquad \tau^{n_o} - t_o t_1^{q_1} \ldots t_n^{q_n}.u = o \quad (u \text{ unit in } A)$$

By taking the generic point $s_o \in D_o$ we see that this equation is irreducible over $\underline{O}_{S,s_o}$ as equation in τ (use for instance Eisenstein's criterium). It follows that T, hence Y, is irreducible. In order to see that $Y \in \mathrm{Rev}^D(S)$ we have only to check condition 5) of definition 2.2.2. For the function field R(Y) of Y we have

$$R(Y) = R(T) = R(S)(\tau) ,$$

where τ is given by the above equation (*). We must show that the extension R(Y) of R(S) is tame in the maximal points of the divisors

D_i. However, this is well known, see for instance [7], proposition 3.4.5 (the fields there are complete with respect to the valuation but by 2.1.4 this makes no difference). Since R(T)= R(Y) is Galois over R(S), with group μ_{n_o}, we have a Galois object Y∈$\mathrm{Rev}^D(S)$. Finally, locally over the generic point of S_o the T and Y are isomorphic (T is normal by 1.7.2 v). Since μ_{n_o} is the inertia group for the points of T_{ξ_o}, the same is true for Y_{ξ_o}. This completes the proof of step I.

<u>Step II</u>. Let $\mathcal{S}$ =Spf A as before and Spf $A_1 \subset \mathcal{S}$ an affine open piece. Construct $\mathcal{Y}_1$ similar as $\mathcal{Y}$ above, we want to see that

$$\mathcal{Y}_1 = \mathcal{Y} \times_{\mathrm{Spf}\ A} \mathrm{Spf}\ A_1 = \mathcal{Y}\big|\ \mathrm{Spf}\ A_1 .$$

The proof of this is entirely similar as the proof of the existence of fibre products in 4.2.2 and depends on the fact that the base change

$$\mathrm{Spec}\ A \leftarrow \mathrm{Spec}\ A_1$$

fulfills the conditions of 2.3.6.

<u>Step III</u>. General case. Follows immediately from step I and II.

<u>7.2.6. Proof of 7.2.4.</u> We write from now on n instead of n_o (n invertible on S_o). From the Kummer sequence in the étale topology of S_o

$$o \to \mu_n \to \mathbb{G}_{m,S_o} \xrightarrow{n} \mathbb{G}_{m,S_o} \to o \quad ,$$

we get the following commutative diagram with exact rows:

$$\begin{array}{ccccccc} \cdots \to & \mathrm{Pic}(S_o) & \xrightarrow{n} & \mathrm{Pic}(S_o) & \xrightarrow{\partial} & H^2(S_o,\mu_n) & \to \cdots \\ & \downarrow \lambda & & \downarrow \lambda & & \downarrow \nu & \\ \cdots \to & \mathrm{Pic}(U_o) & \xrightarrow{n} & \mathrm{Pic}(U_o) & \xrightarrow{\partial} & H^2(U_o,\mu_n) & \to \cdots \end{array}$$

<u>We claim: λ is surjective and ker(λ) is generated by the classes of $D_{i,o}$ (i≠o)</u>

In order to see this we first note that $\mathrm{Div}(S_o) \to \mathrm{Pic}(S_o)$ (resp.

$\mathrm{Div}(U_o) \to \mathrm{Pic}(U_o))$ is surjective, by EGA IV 21.3.4 b, because S_o is reduced. Next on the regular scheme S_o the map $\mathrm{Div}(S_o) \to \mathrm{Div}(U_o)$ is surjective and the kernel is generated by the $D_{i,o}$ $(i \neq o)$ (EGA IV 21.6.9). From these two remarks (and the fact that the kernel from $\mathrm{Div}(-) \to \mathrm{Pic}(-)$ is given by the principal divisors) follows the above claim.

Now let $\alpha \in \mathrm{Pic}(S_o)$ be the <u>self-intersection class</u> of D_o in S_o, i.e., α is the image of class (D_o) by the mapping $\mathrm{Pic}(\mathcal{S}) \to \mathrm{Pic}(S_o)$. The assumption $c'^{(n)}(D_o) = o$ in $H^2(U_o, \mu_n)$ means that $\nu.\partial(\alpha) = o$, hence $\lambda(\alpha) = n\bar{\gamma}$ with $\bar{\gamma} \in \mathrm{Pic}(U_o)$. Let $\gamma \in \mathrm{Pic}(S_o)$ be such that $\lambda(\gamma) = \bar{\gamma}$. Then we have $(\alpha - n\gamma) \in \ker(\lambda)$ i.e.,

$$\alpha - n\gamma = \sum_{i \neq o} (-q_i) \text{ class } (D_{i,o}) .$$

We can assume, adding suitable multiples of n and changing γ, that the q_i are <u>positive</u> integers. Therefore we have:

$$(*) \qquad \left\{\alpha + \sum_{i \neq o} q_i \text{ class } (D_{i,o})\right\} \in n\,\mathrm{Pic}(S_o)$$

with <u>positive integers</u> q_i . Consider now <u>on $\mathcal{S}$ the positive divisor</u>

$$(**) \qquad \bar{D} = D_o + \sum_{i \neq o} q_i D_i \quad .$$

By 6.2.3 lemma 7.2.2 is proved as soon as we see that the element

$$c^{(n)}(\bar{D}) \in H^2(S_o, \mu_{n,\mathcal{S}})$$

is zero, because then $C^{(n)}(\bar{D})(\mathcal{S}) \neq \emptyset$.

We have a natural homomorphism $\mathcal{G}_{m,\mathcal{S}} \to \mathbb{G}_{m,S_o}$ and $\mu_{n,\mathcal{S}} \xrightarrow{\sim} \mu_{n,S_o}$.
by 6.1.3. Using the Kummer sequence (6.1.6 c) :

$$o \to \mu_{n,\mathcal{S}} \to \mathcal{G}_{m,\mathcal{S}} \xrightarrow{n} \mathcal{G}_{m,\mathcal{S}} \to o$$

for the étale topology on S_o we get a commutative diagram of exact sequences (see also 6.1.6 d)

$$\begin{array}{ccccccc}
\to & \mathrm{Pic}(\mathcal{S}) & \xrightarrow{n} & \mathrm{Pic}(\mathcal{S}) & \to & H^2(S_o, \mu_{n,\mathcal{S}}) & \to \cdots \\
 & \downarrow & & \downarrow & & \downarrow\wr & \\
\to & \mathrm{Pic}(S_o) & \xrightarrow{n} & \mathrm{Pic}(S_o) & \to & H^2(S_o, \mu_{n,S_o}) & \to \cdots
\end{array}$$

The fact that $c^{(n)}(\bar{D}) = o$ follows from the isomorphism of the last columm and from (*) and (**). This completes the proof of 7.2.4 and 7.2.2.

7.3. Final result and examples

7.3.1. From 5.1.11 and 7.2.2 we get:

Theorem 7.3.1. Let $\mathcal{S}$ be a connected, locally noetherian regular formal scheme and $(D_i)_{i=0,1\ldots,r}$ a set of divisors on $\mathcal{S}$ with the following properties:

i) the divisors are regular and have normal crossings on $\mathcal{S}$,

ii) the Ideal $\mathcal{Y}(D_o)$ is an Ideal of definition for $\mathcal{S}$,

iii) $D_i \cap D_j = \emptyset$ for $i \neq j$, $i \neq o, j \neq o$.

Under these assumptions and with the divisors (see 5.1.1)

$$D = \sum_i D_i \quad \text{and} \quad D'_o = \sum_{i \neq o} D_{i,o} \quad ,$$

we have an exact sequence

$$\pi_2^{D'_o}(U_o, \xi_o) \xrightarrow{\beta} \mu^t(\xi_o) \xrightarrow{j} \pi_1^D(\mathcal{S}, \xi^*) \to \pi_1^{D'_o}(S_o, \xi_o) \to 1 \ .$$

Here j is the homomorphism described in 5.1.11 (or to be precise: the homomorphism of 5.1.11 followed by the inclusion of K into $\pi_1^D(\mathcal{S}, \xi^*)$) and β is the homomorphism of 7.2.1; β is determined by the self-intersection class of D_o in $\mathcal{S}$.

Remark 7.3.2. The above exact sequence should be compared with part of the exact sequence of homotopy groups obtained, in the theory of differentiable manifolds say, from the normal sphere bundle of S_o in $\mathcal{S}$.

7.3.3. Examples. Let the assumptions and notations be as in 7.3.1 and assume moreover that S_o is an algebraic curve, smooth over a separably closed field k ; let $g = g(S_o)$ be the genus of S_o .

Case a: $g \neq o$ and $I = \{o\}$, i.e., $D = D_o$ and $U_o = S_o$.

Now if $\mathrm{char}(k) \nmid n$ then:

$$H^2(\tilde{S}_o, \mu_n) = o$$

In order to see this let S_1 be an irreducible étale covering of S_o, then $g(S_1) > o$ and S_1 has an irreducible étale covering S_2 of degree n over S_1. Now

$$H^2(S_i, \mu_n) \simeq \mathrm{Pic}(S_i)/n\,\mathrm{Pic}(S_i) \simeq \mathbb{Z}/n\mathbb{Z} \quad (i = 1,2)$$

(see SGA 4 IX 4,7) and by the isomorphism the map

$$H^2(S_1, \mu_n) \to H^2(S_2, \mu_n)$$

corresponds with multiplication by n, therefore every element of $H^2(S_1, \mu_n)$ is killed in $H^2(\tilde{S}, \mu_n) = \varinjlim_{\alpha} H^2(S_\alpha, \mu_n)$. By 7.3.1 we have therefore the exact sequence

$$o \to \mu^t(\xi_o) \xrightarrow{j} \pi_1^D(\mathcal{S}, \xi^*) \to \pi_1(S_o, \xi_o) \to 1$$

and the extension is determined by the self-intersection number of S_o in $\mathcal{S}$. I.e., if d= class $(S_o, S_o) \in \mathrm{Pic}(S_o)$, then the extension corresponds with the image of d in

$$H^2(\pi_1(S_o), \mu^t) \rightrightarrows H^2(S_o, \mu^t) \simeq \hat{\mathbb{Z}}^t$$

where

$$\hat{\mathbb{Z}}^t = \varprojlim_n \mathbb{Z}/n\mathbb{Z} ,$$

with n <u>subject to the condition</u> $\mathrm{char}(k) \nmid n$.

<u>Case b</u>: g= o and $I = \{o\}$, i.e., $U_o = S_o \simeq P^1$ (projective line). Then $\tilde{S}_o^t = S_o$, $H^2(S_o, \mu_n) \simeq \mathbb{Z}/n\mathbb{Z}$, and $\pi_2^{D'_o}(S_o, \xi_o) \simeq \hat{\mathbb{Z}}^t$ (as above). Let again d= class (S_o, S_o) be the self- intersection number of S_o in $\mathcal{S}$, then since $\mathrm{Pic}(S_o) \simeq \mathbb{Z}$, d is a natural number. Put $d' = d.p^e$ ($p = \mathrm{char}.k$, $p \nmid d'$). Identifying $\mu^t(\xi_o)$ with $\hat{\mathbb{Z}}^t$ we have, with the notation of 7.3.1, that

$$\mu^t(\xi_o)/\mathrm{Image}\ (\beta) \simeq \mathbb{Z}/d'\mathbb{Z}$$

i.e.,

$$\pi_1^D(\mathcal{S}, \xi^*) \simeq \mathbb{Z}/d'\mathbb{Z} .$$

<u>Case c</u>: $I = \{o, 1, .., r\}$, $r \neq o$, g arbitrary. In this case U_o is a

curve non-proper over k. Then we have (SGA 4):

$$H^2(U_o,\underline{F}) = o$$

with

$\underline{F}$ torsion, prime to p= char(k)

and also

$$H^2(\tilde{U}_o^t,\underline{F}) = o$$

with

$\underline{F}$ torsion, prime to p= char(k)

Hence $\pi_2(U_o,\xi_o) = o$ and since $H^2(U_o,\mu^t) = o$ we have that $\pi_1^D(\mathcal{S},\xi^*)$ is a semi-direct product:

$$o \to \mu^t(\xi_o) \longrightarrow \pi_1^D(\mathcal{S},\xi^*) \to \pi_1^{D'_o}(S_o,\xi_o) \to 1 \quad .$$

§8. Descent of tamely ramified coverings

8.1. Descent of Modules and Algebras on formal schemes

8.1.1. In section 8.1 the assumptions are as follows: $\mathcal{S}$ is a locally noetherian formal scheme and $\mathcal{J}$ an Ideal of definition for $\mathcal{S}$. Let $\mathcal{J}_\alpha$ ($\alpha = 1,2$) be two coherent Ideals on $\mathcal{S}$ fulfilling the conditions:

i) $$\mathcal{J}_1 \cap \mathcal{J}_2 = \mathcal{J},$$

ii) $$\mathcal{J}_1^n \cap \mathcal{J}_2^n \text{ converges to } 0 \text{ if } n \to \infty .$$

This second condition means that for every positive integer q there exists an integer $n(q) = n$ such that

$$\mathcal{J}_1^n \cap \mathcal{J}_2^n \subset \mathcal{J}^q .$$

Since by i) the $\mathcal{J}^n \subset \mathcal{J}_1^n \cap \mathcal{J}_2^n$, we have that the system $\{\mathcal{J}_1^n \cap \mathcal{J}_2^n\}_{n\in \mathbf{Z}_+}$ is cofinal in the system $\{\mathcal{J}^n\}_{n\in\mathbf{Z}_+}$.

8.1.2. Remark: a) The above conditions are fulfilled in case $\mathcal{S}$ is regular and if we take for every $s \in \mathcal{S}$

$$\mathcal{J}_s = (t_1\ t_2).\underline{O}_{\mathcal{S},s} ,$$

$$\mathcal{J}_\alpha = (t_\alpha).\underline{O}_{\mathcal{S},s} , \qquad (\alpha = 1,2)$$

with t_α ($\alpha = 1,2$) either units or part of a regular system of parameters at s and $\mathcal{J}_1 \cap \mathcal{J}_2 = \mathcal{J}$ an Ideal of definition for $\mathcal{S}$.

b) Let $f: \mathcal{S}' \to \mathcal{S}$ be a flat, adic morphism of locally noetherian formal schemes and $\mathcal{J} = \mathcal{J}_0, \mathcal{J}_1, \mathcal{J}_2$ coherent Ideals on $\mathcal{S}$ having the properties i) and ii) of 8.1.1. Then the Ideals $\mathcal{J}'_\alpha = \mathcal{J}_\alpha . \underline{O}_{\mathcal{S}'}$ ($\alpha = 0,1,2$) have the same properties.

Proof: This is a local question. Let $\mathcal{S} = \mathrm{Spf}\ A$ and $\mathcal{S}' = \mathrm{Spf}\ A'$, then A' is a flat A-algebra. Let J_α (resp. J'_α) be the ideals of A (resp. A') in question. The assertion follows from the fact that for a flat A-algebra A' we have ([3],I, §2, Prop. 6):

$$J_1'^n \cap J_2'^n = J_1^n.A' \cap J_2^n.A' = (J_1^n \cap J_2^n).A' .$$

<u>8.1.3.</u> Put $\mathcal{J}_3 = \mathcal{J}_1 + \mathcal{J}_2$; as a matter of convenient notation write also $\mathcal{J} = \mathcal{J}_o$.

Consider the formal schemes

$$\mathcal{X}_\alpha = (V(\mathcal{J}_\alpha), \varprojlim \underline{O}_{\mathcal{X}}/\mathcal{J}_\alpha^{n+1}) \qquad (\alpha = o,1,2,3) ,$$

i.e., $\mathcal{X}_o = \mathcal{X}$ and $\mathcal{X}_\alpha = \mathcal{X}_{/V(\mathcal{J}_\alpha)}$ $(\alpha = 1,2,3)$ is obtained by starting with the <u>formal</u> scheme $\mathcal{X}$ and <u>completing along</u> $V(\mathcal{J}_\alpha)$. Clearly we have a commutative diagram:

$$\begin{array}{ccccc} & & \mathcal{X}_1 & & \\ & \swarrow^{\lambda_1} & & \nwarrow^{\mu_1} & \\ \mathcal{X} & & & & \mathcal{X}_3 \\ & \nwarrow_{\lambda_2} & & \swarrow_{\mu_2} & \\ & & \mathcal{X}_2 & & \end{array}$$

<u>8.1.4.</u> Let $\mathcal{E}_\alpha$ $(\alpha = o,1,2,3)$ denote the category of <u>coherent</u>, <u>flat</u> $\underline{O}_{\mathcal{X}_\alpha}$-Modules; write also $\mathcal{E} = \mathcal{E}_o$. By $\mathcal{E}^*$ we denote the category of triples $(\mathcal{G}_1, \mathcal{G}_2, \rho)$ with $\mathcal{G}_\alpha \in Ob(\mathcal{E}_\alpha)$ $(\alpha = 1,2)$ and $\rho : \mu_1^*(\mathcal{G}_1) \xrightarrow{\sim} \mu_2^*(\mathcal{G}_2)$ an isomorphism in $\mathcal{E}_3$.

<u>Theorem 8.1.5.</u> Let $\phi : \mathcal{E} \to \mathcal{E}^*$ be the functor defined by

$$\mathcal{F} \rightsquigarrow (\lambda_1^*(\mathcal{F}), \lambda_2^*(\mathcal{F}), \quad \text{can.}: \mu_1^*.\lambda_1^*(\mathcal{F}) \xrightarrow{\sim} \mu_2^*.\lambda_2^*(\mathcal{F})) \text{ with } \mathcal{F} \in Ob(\mathcal{E}).$$

Then ϕ is an <u>equivalence of categories</u>.

<u>Complement:</u> The same is true if we take <u>coherent</u>, <u>flat</u> $\underline{O}_{\mathcal{X}_\alpha}$-Algebras (instead of Modules).

<u>8.1.6.</u> For the proof we use the following lemma:

<u>Lemma 8.1.6:</u> Let A be a noetherian ring, $I_\alpha (\alpha = 1,2)$ ideals in A such

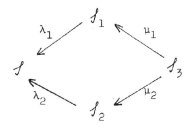

that $I_1 \cap I_2 = (0)$. Put $I_3 = I_1+I_2$ and $A_\alpha = A/I_\alpha$ ($\alpha=1,2,3$). Let E_α denote the category of <u>flat</u> A_α-modules of finite type ($\alpha=0,1\ldots,3$ with $A_0= A$) and E^* denotes the category of triples (M_1, M_2, ρ) with $M_\alpha \in Ob(E_\alpha)$, $\alpha= 1,2$ and

$$\rho : M_1 \otimes_{A_1} A_3 \rightarrow M_2 \otimes_{A_2} A_3 .$$

Consider the functor ϕ: $E= E_0 \rightarrow E^*$ defined for $M\in Ob(E)$ by

$$M \rightsquigarrow \phi(M)= (M \otimes_A A_1, M \otimes_A A_2, \text{can.})$$

Then ϕ is an equivalence of categories.

<u>Proof:</u> a) <u>Faithfulness</u>. Given $\varphi,\psi: M \rightarrow M'$ in E such that $\varphi \otimes 1_{A_\alpha} = \psi \otimes 1_{A_\alpha}$ ($\alpha=1,2$). To prove: $\varphi= \psi$. The above conditions mean

that

$$\varphi(m) \equiv \psi(m) \quad \text{mod } I_\alpha . M' \quad (\alpha=1,2) \quad \text{for all } m\in M.$$

We have by assumption $I_1 \cap I_2= (o)$; since M' is <u>flat</u> this implies ([3]; chap I §2, Prop.6)

$$I_1 .M' \cap I_2 .M' = (I_1 \cap I_2).M' = (o)$$

Hence $\varphi(m)= \psi(m)$ for all $m\in M$.

b) ϕ <u>is fully faithful</u>. Write $M_\alpha = M/I_\alpha M$ and similar for M' ($\alpha=1,2$).

Given

$$\varphi_\alpha : M_\alpha \rightarrow M'_\alpha \quad (\alpha= 1,2)$$

such that

(*) $$\varphi_1 \otimes_{A_1} 1_{A_3} = \varphi_2 \otimes_{A_2} 1_{A_3}$$

we want to find

$$\varphi: M \rightarrow M'$$

such that

$$\varphi \otimes_A 1_{A_\alpha} = \varphi_\alpha .$$

Let $m\in M$, take $x_\alpha \in M'$ such that

$$x_\alpha \equiv \varphi_\alpha(m) \quad \text{mod } I_\alpha .M' \quad (\alpha= 1,2).$$

The condition (*) means that

$$x_1 \equiv x_2 \mod (I_1+I_2).M' \quad .$$

Therefore there exists $m' \in M'$ such that

$$m' \equiv x_\alpha \quad (\text{mod } I_\alpha.M')$$

and such m' is unique modulo $I_1.M' \cap I_2.M'$. Again using the flatness of M' over A we have $I_1.M' \cap I_2.M' = (I_1 \cap I_2).M' = (o)$; i.e., m' is unique. Define $\varphi(m) = m'$ then φ has all the required properties.

c) ϕ <u>is an equivalence</u>. There are given flat A_α-modules M_α which are of finite type and we have a A_3-isomorphism

$$\rho : M_1 \otimes_{A_1} A_3 \xrightarrow{\sim} M_2 \otimes_{A_2} A_3 \; .$$

We are looking for a flat A-module M, of finite type and A_α-isomorphisms

$$\rho_\alpha : M \otimes_A A_\alpha \xrightarrow{\sim} M_\alpha \quad (\alpha = 1,2)$$

such that the following diagram is commutative:

$$\begin{array}{ccc} M_1 \otimes_{A_1} A_3 & \xrightarrow{\rho} & M_2 \otimes_{A_2} A_3 \\ {\scriptstyle \rho_1 \otimes_{A_1} A_3} \nwarrow & & \nearrow {\scriptstyle \rho_2 \otimes_{A_2} A_3} \\ & M \otimes_A A_3 & \end{array}$$

Consider now $S_\alpha = \text{Spec } A_\alpha$ $(\alpha = 0,1,2,3)$, then S_α $(\alpha=1,2,3)$ is a closed subscheme of $S_o = S$. Due to the fact that ϕ is fully faithful, it <u>suffices to prove the existence of (M,ρ_1,ρ_2) locally on S</u> , and we can assume that S is <u>connected</u>. In view of the fact that M_α are flat.A_α-modules of finite type we can assume that they are free, i.e.,

$$M_\alpha \simeq A_\alpha^{(d_\alpha)} \qquad (\alpha = 1,2)$$

From the existence of ρ follows that $d_1 = d_2$ (= d say). Take $M = A^{(d)}$; the problem is now to construct the ρ_α $(\alpha = 1,2)$ compatible with ρ. Now ρ is given by a d-by d-matrix

$$(\sigma_{ij}) \qquad \sigma_{ij} \in A_3$$

and $P_3 = \det(\sigma_{ij})$ is a unit in A_3. Let $s \in S$. In view of the fact that the problem of constructing ρ_α ($\alpha = 1,2$) is local on S, there is only a problem if $s \in S_1 \cap S_2$, i.e., $s \in S_3$. Lift the elements σ_{ij} from A_3 to elements τ_{ij} in A_2.

From the fact that $P_3 = \det(\sigma_{ij})$ is a unit in A_3 we have that $P_2 = \det(\tau_{ij})$ is a <u>unit at s</u> in A_2, hence locally in Spec A_2 a unit. Again, without loss of generality, we can assume that P_2 is a unit in A_2. Then take

$$\rho_1 = \text{can}: A^{(d)} \otimes_A A_1 \xrightarrow{\sim} A_1^{(d)}$$

and for ρ_2 we take the isomorphism

$$\rho_2 : A^{(d)} \otimes_A A_2 \xrightarrow{\sim} A_2^{(d)}$$

determined by the matrix (τ_{ij}).

<u>8.1.7. Proof of 8.1.5</u>. Proceeding as in 8.1.6, i.e., proving first that ϕ is faithful, next ϕ is fully faithful and finally ϕ is an equivalence, we see that the assertions are local on $\mathcal{S}$. Therefore we can assume that $\mathcal{S}$ = Spf A, with A a J-adic noetherian ring. Let J_α ($\alpha = 1,2$) be the ideals corresponding with $\mathcal{J}_\alpha$. Note also that $(J_1 + J_2)^{2n} \subset J_1^n + J_2^n$. Using EGA 0_I 7.2.10 for the fully faithfulness, next EGA 0_I 7.2.9 for the equivalence and [3], Alg. Comm. III, §3, Th.1 for the flatness and the assumption $J_1^n \cap J_2^n \to 0$ for $n \to \infty$, we see that we can replace A by $A/(J_1^n \cap J_2^n)$, the J_α-completion A_α by A/J_α^n ($\alpha = 1,2$) and the J_3-completion A_3 by $A/(J_1^n + J_2^n)$. However, then the assertion reduces to lemma 8.1.6.

The complement about Algebras is obvious.

<u>8.2. Descent of tamely ramified coverings</u>

<u>8.2.1.</u> For the moment the assumptions are the same as in 8.1.1, i.e., we have on $\mathcal{S}$ coherent ideals $\mathcal{J}$, $\mathcal{J}_1$, $\mathcal{J}_2$ with the properties 8.1.1 i)

and ii). Assume moreover that $\mathcal{S}$ is regular and that $(D_i)_{i\in I}$ is a locally finite set of regular divisors with normal crossings on $\mathcal{S}$ (later on in 8.2.6 the Ideals and the divisors will be related to each other) . Introduce as in 8.1.3 the formal schemes $\mathcal{S}_\alpha$ ($\alpha=0,1,2,3$; $\mathcal{S}_0=\mathcal{S}$) and the morphisms $\lambda_\alpha,\mu_\alpha$ ($\alpha=1,2$).

Lemma 8.2.2. The $\mathcal{S}_\alpha$ are regular formal schemes and the inverse images $\left\{\lambda_\alpha^*(D_i)\right\}_{i\in I}$ (resp. $\left\{\mu_\alpha^*\cdot\lambda_\alpha^*(D_i)\right\}_{i\in I}$) are regular divisors with normal crossings on the formal schemes in question.

Note: By abuse of language we denote these inverse images by the same letter D_i ; similarly we use the letter $D=\sum_{i\in I} D_i$ on all four formal schemes.

Proof: The question is local; the lemma follows from 4.1.4 and 3.1.5.

Corollary 8.2.3. Let $\mathcal{X}\in \mathrm{Rev}^D(\mathcal{S})$; consider the inverse image

$$\lambda_\alpha^*(\mathcal{X})=\mathcal{X}/_{V(\mathcal{J}_\alpha)}=\mathcal{X}\times_{\mathcal{S}}\mathcal{S}_\alpha \qquad (\alpha=1,2)$$

Then $\lambda_\alpha^*(\mathcal{X})$ is in $\mathrm{Rev}^D(\mathcal{S}_\alpha)$ for $\alpha=1,2$.

Similarly: the inverse image under μ_α of an object of $\mathrm{Rev}^D(\mathcal{S}_\alpha)$ ($\alpha=1,2$) is in $\mathrm{Rev}^D(\mathcal{S}_3)$.

Proof: Again the question is local on $\mathcal{S}$. Using the relation between usual and formal tamely ramified coverings (4.1.3), the corollary follows from 8.2.2 and 4.1.4.

8.2.4. Let $\mathcal{R}$ denote the category of triples $(\mathcal{X}_1,\mathcal{X}_2,\rho)$ with $\mathcal{X}_\alpha\in\mathrm{Rev}^D(\mathcal{S}_\alpha)$ ($\alpha=1,2$) and ρ an isomorphism in $\mathrm{Rev}^D(\mathcal{S}_3)$:

$$\rho : \mu_1^*(\mathcal{X}_1) \xrightarrow{\sim} \mu_2^*(\mathcal{X}_2) .$$

Theorem 8.2.5. Consider the functor

$$\phi.\ \mathrm{Rev}^D(\mathcal{Y}) \to \mathcal{R}$$

defined by

$$\phi(\mathfrak{X}) = (\lambda_1^*(\mathfrak{X}),\ \lambda_2^*(\mathfrak{X}),\ \text{can.}) \qquad (\text{with } \mathfrak{X} \in \mathrm{Rev}^D(\mathcal{Y}))\ .$$

This is an equivalence of categories.

+) see page 118.

Proof: The fact that ϕ is well-defined follows from 8.2.3. Now recall that for $f: \mathfrak{X} \to \mathcal{Y}$ in $\mathrm{Rev}^D(\mathcal{Y})$ the $f_*(\underline{O}_{\mathfrak{X}})$ is a coherent, flat $\underline{O}_{\mathcal{Y}}$- Algebra (2.3.5 and 3.1.7). The full faithfulness of ϕ follows from 8.1.5. Next start with $(\mathfrak{X}_1, \mathfrak{X}_2, \rho) \in \mathrm{Ob}(\mathcal{R})$; by 8.1.5 we get a coherent, flat $\underline{O}_{\mathcal{Y}}$-Algebra $\mathcal{B}$, put $\mathfrak{X} = \mathrm{Spf}\ \mathcal{B}$ (3.1.6). We want to show that $\mathfrak{X} \to \mathcal{Y}$ is tamely ramified relative to D. This is a local question on $\mathcal{Y}$; let $\mathcal{Y} = \mathrm{Spf}\ A$, A a noetherian J-adic ring and $\mathfrak{X} = \mathrm{Spf}\ B$. Introduce also S=Spec A and X =Spec B.

Let $s \in \mathcal{Y}$; since $\mathcal{J}_1 \cap \mathcal{J}_2 = \mathcal{J}$ we have either $s \in V(\mathcal{J}_1)$ or $s \in V(\mathcal{J}_2)$ or both. Let $s \in V(\mathcal{J}_1)$, say. Write A' for the J_1-adic completion of A, $S' = \mathrm{Spec}\ A'$ and $X' = X \times_S S'$.

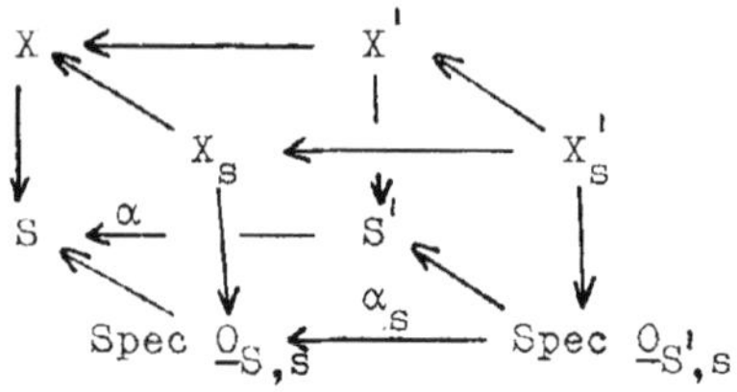

By the usual interplay between the local rings $\underline{O}_{S,s}$ and $\underline{O}_{\mathcal{Y},s}$ (resp. $\underline{O}_{S',s}$ and $\underline{O}_{\mathcal{Y},s}$) it suffices to show by 2.3.6 that $X_s = X \times_S \mathrm{Spec}\ (\underline{O}_{S,s})$ is tamely ramified over Spec $(\underline{O}_{S,s})$ relative to the divisor D (resp. we know by 2.2.9 that X'_s is tamely ramified over Spec $\underline{O}_{S',s}$). We have that α is flat (EGA O_I 7.3.3), hence α_s is flat. The tame ramification of X_s follows then by 2.2.9, applied to α_s, from the tame ramification of X'_s . This completes the proof of 8.2.5.

8.2.6. Application. At this point we drop the assumptions and notations made in 8.2.1.and make the following assumptions instead:

$\mathcal{J}$ is a locally noetherian, regular formal scheme and $(D_i)_{i\in I}$ a finite set of regular divisors with normal crossings on $\mathcal{J}$ such that

$$D_i \cap D_j \cap D_k = \emptyset \qquad (i\neq j,\ i\neq k,\ j\neq k)$$

Put $D = \sum_{i\in I} D_i$ and assume that $\mathcal{J}(D)$ is an Ideal of definition for $\mathcal{J}$.

Introduce the formal schemes $\mathcal{J}_i = \mathcal{J}_{/D_i}$ and $\mathcal{J}_{ij} = \mathcal{J}_{/D_i \cap D_j}$; i.e., if $\mathcal{J}(D_i)$ is the Ideal determined by D_i then we put

$$\mathcal{J}_i = (D_i, \varprojlim_n \mathcal{O}_{\mathcal{J}}/\mathcal{J}(D_i)^{n+1})$$

and

$$\mathcal{J}_{ij} = (D_i \cap D_j, \varprojlim_n \mathcal{O}_{\mathcal{J}}/\{\mathcal{J}(D_i) + \mathcal{J}(D_j)\}^{n+1}) .$$

Furthermore $\lambda_i : \mathcal{J}_i \to \mathcal{J}$, resp. $\lambda_{i,ij} : \mathcal{J}_{ij} \to \mathcal{J}_i$, are the natural morphisms. By lemma 8.2.2 the inverse images $\lambda_i^*(D_k)$ exist and are regular divisors with normal crossings on $\mathcal{J}_i$ (i fixed, $k\in I$; note that $k = i$ occurs also!), we denote these inverse images by the same letters D_k. A similar remark holds for the $\lambda_{i,ij}^*(D_k)$.

Finally, let $\mathcal{R}$ denote the category of systems $\{\mathfrak{X}_i, \rho_{ij}\}_{i,j\in I}$ with $\mathfrak{X}_i \in \mathrm{Rev}^D(\mathcal{J}_i)$ and ρ_{ij} isomorphisms in $\mathrm{Rev}^D(\mathcal{J}_{ij})$ (cf.8.2.3) :

$$\rho_{ij} : \lambda_{j,ij}^*(\mathfrak{X}_j) \overset{\sim}{\rightarrow} \lambda_{i,ij}^*(\mathfrak{X}_i) .$$

Theorem 8.2.7. With the above assumptions and notations, consider the functor

$$\phi : \mathrm{Rev}^D(\mathcal{J}) \to \mathcal{R}$$

defined by

$$\phi(\mathfrak{X}) = \left\{\lambda_i^*(\mathfrak{X}),\ \mathrm{can.}\right\} \qquad (\mathfrak{X}\in \mathrm{Rev}^D(\mathcal{J})).$$

This functor is an equivalence. $^{+}$)

$^{+}$) see page 118.

Proof: By 8.2.3 the functor is well-defined. Apply induction to

Card (I).*) Take $i_o \in I$, put $I' = I - \{i_o\}$ and put $D' = \sum_{i \neq i_o} D_i$. Introduce the Ideals $\mathcal{J}_1 = \mathcal{J}(D_{i_o})$, $\mathcal{J}_2 = \mathcal{J}(D')$. Since $\mathcal{S}$ is regular the Ideals $\mathcal{J}_1$ and $\mathcal{J}_2$ fulfil the conditions of 8.1.1 (see 8.1.2); in particular $\mathcal{J}_1 \cap \mathcal{J}_2 = \mathcal{J}(D)$. Finally put $\mathcal{J}_3 = \mathcal{J}_1 + \mathcal{J}_2$.

Let $\mathcal{T}_i = \mathcal{S}_{/V(\mathcal{J}_i)}$, i= 1,2,3 then $\mathcal{T}_1 = \mathcal{S}_{i_o}$. Let $\sigma_\alpha : \mathcal{T}_\alpha \to \mathcal{S}$ (resp. $\tau_\alpha : \mathcal{T}_3 \to \mathcal{T}_\alpha$) ($\alpha$= 1,2) be the natural morphisms.

As remarked above the $\mathcal{S}$, $\mathcal{J}_1$ and $\mathcal{J}_2$ fulfil the conditions of theorem 8.2.5. Therefore the category $\mathrm{Rev}^D(\mathcal{S})$ is equivalent with the category $\mathcal{R}^*$ of triples $(\mathcal{X}_{i_o}, \mathcal{X}', \rho)$ with $\mathcal{X}_{i_o} \in \mathrm{Rev}^D(\mathcal{S}_{i_o})$, $\mathcal{X}' \in \mathrm{Rev}^D(\mathcal{T}_2)$ and $\rho : \tau_1^*(\mathcal{X}_{i_o}) \xrightarrow{\sim} \tau_2^*(\mathcal{X}')$ an isomorphism in $\mathrm{Rev}^D(\mathcal{T}_3)$.

Next remark that, due to the assumption $D_i \cap D_j \cap D_k = \emptyset$ for $i \neq j$, $i \neq k$, $j \neq k$, we have for $s \in D_{i_o} \cap D_i$ that $\mathcal{J}_3 = \mathcal{J}(D_{i_o}) + \mathcal{J}(D_i)$ in a Zariski neighbourhood of s. From this remark and the induction assumption applied on $\mathcal{T}_2$, we have that the category $\mathcal{R}^*$ is equivalent with the category $\mathcal{R}$; q.e.d.

*Footnote:

In order to be able to apply an induction, we must consider the following slightly more general situation. Let $D = \sum_i D_i + \sum_j D_j^*$ with $(D_i, D_j^*)_{i,j}$ regular divisor with normal crossings on $\mathcal{S}$, such that every triple has empty intersection. Let $\mathcal{J}(\sum_i D_i)$ be the Ideal of definition for $\mathcal{S}$. I.e., the D_j^* are in the game as far as the tame ramification is concerned, but they are not used for taking completions. Introduce $\mathcal{S}_i = \mathcal{S}_{/D_i}$ etc... Theorem 8.2.7 can be formulated for this situation.

+) In 8.2.5 and 8.2.7 we have used the abbreviation "can." for the obvious canonical isomorphisms; for instance in 8.2.7 we have:

$$\text{can.} : (\lambda_i . \lambda_{i,ij})^*(\mathcal{X}) \xrightarrow{\sim} (\lambda_j . \lambda_{j,ij})^*(\mathcal{X}) .$$

8.3. Reformulation of the previous results in terms of the fundamental groups

8.3.1. Assumptions. $\mathcal{S}$ is a connected, regular locally noetherian formal scheme and $(D_i)_{i\in I}$ a finite set of regular, connected divisors with normal crossings on $\mathcal{S}$ such that

$$D_i \cap D_j \cap D_k = \emptyset \qquad (i \neq j,\ i \neq k,\ j \neq k)$$

and, if $D = \sum_i D_i$, then $\mathcal{J}(D)$ is an Ideal of definition for $\mathcal{S}$.

Notations: Put $\mathcal{S}_i = \mathcal{S}_{/D_i}$ and $\mathcal{S}_{ij} = \mathcal{S}_{/D_i \cap D_j}$; let $\lambda_i : \mathcal{S}_i \to \mathcal{S}$ (resp. $\lambda_{i,ij} : \mathcal{S}_{ij} \to \mathcal{S}_i$) be the natural morphisms. Finally, by lemma 8.2.2 the inverse images of the divisors under these morphisms exist and are regular divisors with normal crossings; they are denoted by the same letters as on $\mathcal{S}$. Note that the $\mathcal{S}_i$ are connected.

Fundamental groups: Let E''_{ij} be the set of connected components of $\mathcal{S}_{ij}$, put $E'' = \bigcup_{i,j} E''_{ij}$. In order to have a convenient notation we introduce also the set E' of (connected!) formal schemes $\mathcal{S}_i$ (i.e., $E' = I$) and $s'_i \in E'$ denotes the formal scheme $\mathcal{S}_i$. Sometimes we write shortly $s' \in E'$ (resp. $s'' \in E''$) if it is not necessary to emphasis which $\mathcal{S}_i$ (resp. which connected component of $\mathcal{S}_{ij}$) we take. For every $s' \in E'$ (resp. $s'' \in E''$) we choose a base point $\xi_{s'}$ (resp. $\xi_{s''}$). Put:

$$\pi_{s'} = \pi^D(s', \xi_{s'}) \qquad (\text{resp. } \pi_{s''} = \pi^D(s'', \xi_{s''})).$$

Finally, every $s''_{ij} \in E''_{ij}$ determines an element $s'_i \in E'$ and $s'_j \in E'$ (here again $s'_i(s''_{ij})$ would be a more precise, but also a more cumbersome notation) . We choose a path (cf.4.2.5):

$$q_i^{s''_{ij}} : \pi_{s''_{ij}} \to \pi_{s'_i} \ ,$$

and similarly for j instead of i. For simplicity we often write $q_i^{s''}$ (resp. $q_j^{s''}$) instead of $q_i^{s''_{ij}}$ (resp. $q_j^{s''_{ij}}$) .

8.3.2. Given the $\pi_{s'}$ $(s' \in E')$ and the $q_i^{s''}$ and $q_j^{s''}$ $(s'' \in E'')$, we want to determine the $\pi = \pi^D(\mathcal{S}, \xi)$ for a suitable base point ξ. Fix $i_0 \in I$ and $s'_0 \in E_{i_0}$; take the base point $\xi_{s'_0}$, this determines a base point ξ for $\mathcal{S}$. Then $\mathrm{Rev}^D(\mathcal{S})$ is equivalent, by means of the fibre functor $\mathcal{X} \to \mathcal{X}_\xi$, with the Galois category $C(\pi)$. According to 8.2.7 an object A of $C(\pi)$ corresponds uniquely (after a choice of paths from ξ_s to ξ!) with a system of

i) $A'_{s'} \in C(\pi_{s'})$ $(\forall s' \in E')$,

ii) isomorphisms $\varphi_{s''_{ij}} : A'_{s'_i} \to A'_{s'_j}$ $(\forall s'' \in E'')$

fulfilling the conditions

$$\varphi_{s''_{ij}}(q_i^{s''_{ij}}(g).x) = q_j^{s''_{ij}}(g).\varphi_{s''_{ij}}(x) \qquad (\forall x \in A'_{s'_i}, \forall g \in \pi_{s''_{ij}}) .$$

8.3.3. The sets (E', E'') constitute a graph with the elements of E' as vertices.

Due to the connectedness of $\mathcal{S}$, this is a connected graph. It is possible to make a partition

$$E'' = E''_* \cup \tilde{E}'' \qquad \text{with } E''_* \cap \tilde{E}'' = \emptyset$$

such that (E', E''_*) is still a connected graph but without closed circuits (proof: induction on card (E')); we may consider then (E', E''_*) as a tree with s'_0 as "trunk".

8.3.4. Suppose now that we have a system as in 8.3.2. By means of the $\varphi_{s''}$ with $s'' \in E''_*$ we can identify $A'_{s'_i}$ with $A'_{s'_0}$ which we denote by A_0. The $\pi_{s'_i}$ operate now on A_0 and we have bijections

$$g_{s''} : A_0 \to A_0 \qquad (\forall s'' \in E'')$$

satisfying the following conditions

i) $g_{s''} . q_i^{s''}(g) = q_j^{s''}(g) . g_{s''}$ $(s'' \in E''_{ij} \subset E'', \forall g \in \pi_{s''})$,

ii) $g_{s''} = 1$ $(s'' \in E''_*)$

After the identifications we have, by 8.2.6, that the category

$C(\pi)$ is equivalent with the category of finite sets on which the $\pi_{s'}$ ($s' \in E'$) operate continuously and on which the $g_{s''}$ ($s'' \in E''$) operate subject to the above relations. Therefore we have the following theorem and corollary:

Theorem 8.3.5. The fundamental group $\pi = \pi_1^D(\mathcal{S}, \xi)$ is the topological group of Galois type *) generated by the $\pi_{s'}$ ($s' \in E'$) and the $g_{s''}$ ($s'' \in E''$) subject to the relations i) and ii) in 8.3.4 above.

8.3.6. If p is a prime number and G is a (pro)finite group, then we denote by $G^{(p)}$ the profinite group

$$G^{(p)} = \varprojlim_{\alpha} \mathcal{G}_{\alpha} \quad (\mathcal{G}_{\alpha} \text{ a finite quotient of } G \text{ of order prime to } p)$$

Corollary 8.3.6. $\pi^{(p)} = \left\{ \pi^D(\mathcal{S}, \xi) \right\}^{(p)}$ is the topological group of Galois type and of "order prime to p" generated by the $\pi_{s'}^{(p)}$ ($s' \in E'$) and the $g_{s''}$ ($s'' \in E''$) subject to the above relations i) and ii) of 8.3.4.

----

*Footnote:

By this is meant the following (cf. also SGA 1 IX section 5). Consider the group Γ generated by the $\pi_{s'}$ and the $g_{s''}$, subject to the relations i) and ii). The group π is the projective limit of those quotients $\mathcal{G}$ of Γ, for which the order of $\mathcal{G}$ is finite and upon which the action of $\pi_{s'}$ is continuous. In 8.3.6 we require moreover that the order of $\mathcal{G}$ is prime to p.

§9. An application: the fundamental group of the spectrum of a complete local ring, of dimension two, minus a closed set

9.1. Let A be a noetherian local ring with the following properties:

i) A is complete,

ii) A has dimension 2,

iii) A has algebraically closed residue field $k = A/\underline{m}$.

Let $S' = \operatorname{Spec} A$ and E a closed set of S' such that $S = S' - E$ is connected (of particular interest is the case $E = \{\underline{m}\}$). Select a base point ξ_0 in S; in the following this base point is suppressed in the notation. Finally let p be the characteristic of k (p is zero or a prime number). Then we have the following (cf. SGA 2 XIII 3.1.ii):

Theorem 9.2. Let $\pi_1(S) = \pi_1(S, \xi_0)$ be the fundamental group of S and $\pi_1^{(p)}(S)$ the largest profinite quotient of $\pi_1(S)$ "of order prime to p". Then $\pi_1^{(p)}(S)$ is topologically of finite presentation.

Remarks:

a) Recall that

$$\pi_1^{(p)}(S) = \varprojlim \mathcal{G}_\alpha ,$$

with $\mathcal{G}_\alpha$ running through the finite quotients of $\pi_1(S)$, of order prime to p (see 8.3.6). Sometimes we write also $\{\pi_1(S)\}^{(p)}$

b) Recall that a profinite group Γ "of order prime to p" is said to be topologically of finite presentation if:

i) There exists a finite set I of elements $g_i \in \Gamma$ $(i \in I)$ such that the subgroup generated by the g_i in Γ is dense in Γ.

ii) Let $F^{(p)}(I)$ be the free profinite group of order prime to p on the set I, i.e.,

$$F^{(p)}(I) = \varprojlim \mathcal{h}_\alpha$$

with h_α running through the finite quotients of the free group $F(I)$ and of order prime to p. Consider the continuous homomorphism

$$\varphi\colon F^{(p)}(I) \to \Gamma$$

obtained by mapping i to g_i. Then there exists a finite set of elements $r_j \in F^{(p)}(I)$ $(j \in J)$ such that the closed normal subgroup generated by the r_j in $F^{(p)}(I)$ is $\ker(\varphi)$.

9.3. Proof of 9.2. Reduction to the case of an integral local ring still satisfying the conditions of 9.1:

Let $\mathfrak{p}_i$ be the minimal prime ideals belonging to (o) in A; put $A_i = A/\mathfrak{p}_i$, $S'_i =$ Spec A_i, S_i the inverse image of S in S'_i. Consider the natural morphism

$$\varphi\colon S^* = \coprod_i S_i \to S.$$

Assuming the theorem for the S_i, it follows from SGA 1 IX 4.12 and 5.2. that $\pi_1^{(p)}(S)$ itself is topologically finitely generated. In fact this is still true for the fundamental groups (prime to p) of the components of S in case S is not connected.

Next consider $S'_i \times_{S'} S'_j$; this again is a local ring of type considered in 9.1. except that possibly condition ii) is replaced by dimension ≤ 2. From the above remark follows the finite generation (topologically) of the part prime to p of the fundamental groups of the connected components of $S_i \times_S S_j$ in case of dimension 2; in case of dimension 1 we normalize and use [6]. p.75 Cor 2 and 4.

It follows then from SGA 1 IX 5.3 applied to the morphism φ that $\pi_1^{(p)}(S)$ is topologically of finite presentation as soon as the $\pi_1^{(p)}(S_i)$ have this property.

9.4. Reduction to the case of normal local rings (of type 9.1): Starting with integral A we take the normalization A_1 of A.

By EGA 0_{IV} 23.1.5 we have that A_1 is finite over A, hence semi-local, and hence local since A_1 is integral. The argument is now similar to the one used in 9.3.

<u>9.5.</u> From now on we make, besides the conditions of 9.1, the <u>additional assumption</u> that A is <u>normal</u> (of dimension 2; the case of dimension 1, which appears in course of the above reduction, is treated as described in 9.3). The notations are as in 9.1.

We follow the method of SGA 2 XIII §3, which was inaugurated <u>by Mumford</u> [5]. By <u>Abhyankar</u> [1] and [1¹] we have a <u>desingularization</u>

$$f: T' \rightarrow S'$$

of S' with the following properties. The morphism f is <u>proper</u>; the inverse image $f^{-1}(E) = D$ is a closed subscheme which can be considered as a <u>divisor</u>

$$D = \sum_i D_i$$

with $(D_i)_{i \in I}$ a finite set of <u>regular</u> irreducible divisors with <u>normal crossings</u> on T' and with

$$D_i \cap D_j \cap D_q = \emptyset \qquad (i \neq j,\ j \neq q,\ i \neq q)\ .$$

Furthermore T' is a <u>regular</u> scheme and if $T = T' - D$ then, due to the normality of S', the restriction $f|T$ is an isomorphism.

Let $\mathcal{T}' = T'_{/D}$ be the completion of T' along D. We have now the following situation:

$$\begin{array}{ccccc} T & \hookrightarrow & T' & \longleftarrow & \mathcal{T}' \\ \cong \big\downarrow f & & \big\downarrow f & & \\ S & \hookrightarrow & S' & & \end{array}$$

As a matter of notation, if Z is a scheme (resp. a formal scheme), then we denote the category of <u>coverings</u> of Z, i.e., the category of schemes (resp. formal schemes) finite over Z, by Rev(Z).

<u>Lemma 9.6.</u> The natural functor "completion along D":

$$\phi : \mathrm{Rev}(T') \to \mathrm{Rev}(\mathcal{T}')$$

is an equivalence and its restriction (still denoted by the same letter!) to $\mathrm{Rev}^D(T')$ gives an <u>equivalence</u>:

$$\phi : \mathrm{Rev}^D(T') \to \mathrm{Rev}^D(\mathcal{T}').$$

(Note: as usual we have denoted the inverse image of D on $\mathcal{T}'$ by the same letter D).

<u>Proof</u>: The first statement follows from the comparison and existence theorem EGA III 5.1.6, because A is complete. The fact that for $X \in \mathrm{Rev}^D(T')$ the $\phi(X) \in \mathrm{Rev}^D(\mathcal{T}')$ follows from 4.1.4. To see that this restriction is still an equivalence we start with $\mathfrak{X} \in \mathrm{Rev}^D(\mathcal{T}')$, let $X \in \mathrm{Rev}(T')$ be such that $\phi(X) = \mathfrak{X}$. Let $g: X \to T'$ be the structure map, put $g_*(\underline{O}_X) = \underline{B}$, then $\mathfrak{X} = \mathrm{Spf}\ \hat{\underline{B}}$ (where ^ means completion along D). Let $t \in D$; consider the stalk $\underline{B}_t$ (resp. $(\hat{\underline{B}})_t$) of $\underline{B}$ (resp. $\hat{\underline{B}}$) at $t \in T'$ (resp. $t \in \mathcal{T}'$).

By the definition of tame ramification (4.1.2) of $\mathfrak{X}$ we have that the (usual!) scheme $\mathrm{Spec}\ (\hat{\underline{B}})_t$ is normal. Furthermore the diagram

$$\begin{array}{ccc} \mathrm{Spec}\ \underline{B}_t & \xleftarrow{\beta} & \mathrm{Spec}((\hat{\underline{B}})_t) \\ \downarrow & & \downarrow \\ \mathrm{Spec}\ \underline{O}_{T',t} & \xleftarrow{\alpha} & \mathrm{Spec}\ \underline{O}_{\mathcal{T}',t} \end{array}$$

is <u>cartesian</u> (EGA 0_I 7.7.8). From the flatness of α (EGA I 10.8.9) follows the flatness of β. By EGA IV 6.5.4 i) the normality of $\mathrm{Spec}\ (\hat{\underline{B}})_t$ implies the normality of $\mathrm{Spec}\ \underline{B}_t$. Therefore X is normal at all points x with $g(x) \in D$; since both f and g are proper it follows that X is normal at all closed points, hence X is normal. From 4.1.5 follows that X is tamely ramified over an open neighbourhood of D, but then, since f is proper, it follows that $X \in \mathrm{Rev}^D(T')$.

<u>Lemma 9.7.</u> Let $\mathrm{Rev}_{nor,sep}(T')$ denote the full subcategory of $\mathrm{Rev}(T')$, consisting of schemes Y which are <u>normal</u> and for which the function

ring R(Y) is a union of separable extensions of the function field R(S)= R(T'). Then the restriction functor

$$\psi : Rev_{nor,sep}(T') \to Rev_{nor,sep}(T)$$

is an equivalence.

Proof: The restriction functor ψ : Rev(T') → Rev(T) is faithful because T is the complement of a divisor on T'. By EGA II 6.3.9 ψ is fully faithful if we restrict to the full subcategory of normal schemes Y'∈ Rev(T'). It is an equivalence if we require moreover that the function ring of the schemes Y∈Rev(T) is separable over R(S)=R(T). Because starting with Y∈$Rev_{nor,sep}$(T), the normalization Y' of T' in R(Y) is finite over T' ([2], Alg.Comm, V, §1,no 6, cor 1 of prop.18).

Corollary 9.8. The restriction functor

$$\psi : Rev^{D}(T') \to Ret\ Et(S)$$

defines a continuous homomorphism

$$\lambda: \pi_1(S) \to \pi_1^D(T')$$

such that

$$\mu = \lambda^{(p)}: \{\pi_1(S)\}^{(p)} \to \{\pi_1^D(T')\}^{(p)}$$

is an isomorphism (Note: Rev Et(S) is -as usual- the category of étale coverings).

Proof: Note that Rev Et(S)= Rev Et(T). The first assertion follows from general Galois theory (SGA 1 V, §6).
If Y'∈ Rev^D(T') is connected then, due to the normality, it is irreducible, hence Y'|T is connected. Therefore λ,and hence μ, is surjective.

In order to see that μ is injective it suffices to start with a Y∈ Rev Et(S), which is Galois with group Γ prime to p, and to show that the, by 9.7 up to isomorphism unique Y'∈ $Rev_{nor,sep}$(T') with ψ(Y')= Y, is in fact in Rev^D(T').

The only condition which remains to be checked is condition 5 of 2.2.2; however, since $R(Y')$ is Galois with group Γ of order prime to p, this condition is automatically fulfilled by 2.1.3 v.

Corollary 9.9. We have natural isomorphisms

$$\left\{\pi_1^D(\mathcal{G}')\right\}^{(p)} \xrightarrow{\sim} \left\{\pi_1^D(T')\right\}^{(p)} \xleftarrow{\sim} \left\{\pi_1(S)\right\}^{(p)}$$

Proof: Combine 9.8 and 9.6.

9.10. In order to prove that $\left\{\pi_1(S)\right\}^{(p)}$ is topologically of finite presentation it suffices to prove this for $\left\{\pi_1^D(\mathcal{G}')\right\}^{(p)}$. For this, in turn, it suffices to look to the groups $\pi_{s'}$ (resp. $\pi_{s''}$) entering in 8.3.6 and to prove that these are (as far as the part prime to p is concerned) topologically of finite presentation (resp. finite generation).

9.11. Finite presentation of the $\pi_{s'}^{(p)}$. Each s' in 8.3.6 corresponds with an irreducible divisor D_i. This D_i is either a curve over the algebraically closed residue field k or it corresponds with a discrete valuation ring with algebraically closed residue field. In the former case the $\pi_{s'}^{(p)}$ is topologically of finite presentation by the examples of 7.3.3; at this point we use the results of SGA 1 X and XII.
In the latter case we use the exact sequence 7.3.1; this sequence takes here the form (cf. also SGA 4 X 2.2)

$$1 \rightarrow \mu^t \rightarrow \pi_1^D(\mathcal{S}) \rightarrow \pi_1^{D'_o}(S_o) \rightarrow 1 .$$

If $D'_o = \emptyset$ we have $\pi_1(S_o) = o$; if $D'_o \neq \emptyset$ then $\left\{\pi_1^{D'_o}(S_o)\right\}^{(p)}$ is (topologically) cyclic by [6], p.75 cor.2 or cor. 4. Therefore in both cases $\left\{\pi_1^D(S)\right\}^{(p)}$ is topologically of finite presentation.

9.12. Finite generation (in fact: finite presentation) of $\pi_{s''}^{(p)}$. By our assumption on the divisors and from the way the formal scheme $\mathcal{S}''$ is constructed (8.2.6), we have that $\mathcal{S}''$ = Spf B with B a complete local ring, noetherian, of dimension 2, <u>regular</u> and with algebraically closed residue field. By 4.1.3 it suffices to consider $\mathrm{Rev}^D(\mathrm{Spec}\ B)$. Since B is <u>strict hensel</u> (EGA IV 18.5.16), we have by 2.3.4 that such a tamely ramified covering is a disjoint union of generalized Kummer coverings. From this, and from the fact that in Spec B we have $D = D_1 + D_2$ with D_1 and D_2 regular divisors with normal crossings, we get

$$\pi_{s''} \simeq \mu^t \times \mu^t$$

with

$$\mu^t = \varprojlim_n \mu_n \quad (p \nmid n).$$

Hence we have (topologically) finite presentation.

Index terminology

Symbols:

References

[1] S.S. Abhyankar, "Resolution of singularities of arithmetical surfaces", Arithmetical Algebraic Geometry, Harper-Row, New York, 1965, p.111-153.

[1] S.S. Abhyankar, "On valuations centered in a local domain", Am.J. of Math., Vol 78, 1956, p.321-348.

[2] M. Artin, "Grothendieck Topologies", mimeographed notes, Harvard, 1962.

[SGA4] M.Artin and A. Grothendieck, "Séminaire de Géométrie Algébrique", no.4 mimeographed notes, I.H.E.S., Paris, 1963-'64.

[3] N. Bourbaki, "Algèbre Commutative", Hermann, Paris, 1961-'65

[SGA3] M. Demazure and A. Grothendieck, "Schémas en Groupes", Sém. Géom.Alg., Lecture Notes in Math., no.151-153, Springer Verlag, Berlin, 1970.

[4] J. Giraud, "Cohomologie non abélienne", mimeographed notes, Columbia Univ., 1966.

[EGA] A. Grothendieck and J. Dieudonné, "Eléments de Géométrie Algébrique", Publ.Math., I.H.E.S., no.4,8,11, 20,24,28 and 32, Paris, 1960 ff.

[SGA1 and 2] A. Grothendieck, "Séminaire de Géométrie Algébrique", no.1, mimeographed notes, I.H.E.S., Paris, 1960-'61; no.2, North-Holland Publ.Co, Amsterdam, 1968.

[5] D. Mumford, "The topology of normal singularities of an algebraic surface and a criterion for simplicity", Publ.Math., I.H.E.S., no.9, Paris, 1961, p.5-22.

[6] J.P. Serre, "Corps locaux", Hermann, Paris, 1962.

[7] E. Weiss, "Algebraic number theory", Mc.Graw-Hill, New York, 1963.

[8] O. Zariski and P. Samuel, "Commutative Algebra", Vol.I, Van Nostrand, New York, 1958.